Finite Element Analysis

CONTENTS

UNIT – 3 TWO DIMENSIONAL FINITE ELEMENT ANALYSIS

UNIT – 4 DYNAMIC ANALYSIS USING ELEMENT METHOD

UNIT -5 APPLICATIONS IN HEAT TRANSFER &FLUID MECHANICS

TEXT BOOKS

:

1. P.Seshu, "Text Book of Finite Element Analysis", Prentice-Hall of India Pvt. Ltd. New Delhi, 2007.
2. J.N.Reddy, "An Introduction to the Finite Element Method", McGraw-Hill International Editions(Engineering Mechanics Series), 1993.
3. Cook,Robert.D., Plesha,Michael.E & Witt,Robert.J. "Concepts and Applications of Finite Element Analysis",Wiley Student Edition, 2004.
4. Chandrupatla & Belagundu, "Introduction to Finite Elements in Engineering", 3rd Edition, Prentice-Hall of India, Eastern Economy Editions.

FINITE ELEMENT ANALYSIS

OBJECTIVES:

1. **To introduce the concepts of Mathematical Modeling of Engineering Problems.**
2. **To appreciate the use of FEM to a range of Engineering Problems**

UNIT I INTRODUCTION **9**

Historical Background – Mathematical Modeling of field problems in Engineering – Governing Equations –Discrete and continuous models – Boundary, Initial and Eigen Value problems– Weighted Residual Methods – Variational Formulation of Boundary Value Problems – RitzTechnique – Basic concepts of the Finite Element Method.

UNIT II ONE DIMENSIONAL PROBLEM **9**

One Dimensional Second Order Equations – Discretization – Element types- Linear and Higher order Elements – Derivation of Shape functions and Stiffness matrices and force vectors- Assembly of Matrices –Solution of problems from solid mechanics and heat transfer. Longitudinal vibration frequencies and mode shapes. Fourth Order Beam Equation –Transverse deflections and Natural frequencies of beams.

UNIT III TWO DIMENSIONAL SCALAR VARIABLE PROBLEMS **9**

Second Order 2D Equations involving Scalar Variable Functions – Variational formulation –Finite Element formulation – Triangular elements – Shape functions and element matrices and vectors. Application to FieldProblems – Thermal problems – Torsion of Non circular shafts –Quadrilateral elements – Higher Order Elements.

UNIT IV TWO DIMENSIONAL VECTOR VARIABLE PROBLEMS **9**

Equations of elasticity – Plane stress, plane strain and axisymmetric problems – Body forces andtemperature effects – Stress calculations – Plate and shell elements.

UNIT V ISOPARAMETRIC FORMULATION **9**

Natural co-ordinate systems – Isoparametric elements – Shape functions for iso parametric elements – Oneand two dimensions – Serendipity elements – Numerical integration and application to plane stress problems – Matrix solution techniques – Solutions Techniques to Dynamic problems – Introduction to Analysis Software.

OUTCOMES:

Upon completion of this course, the students can able to understand different mathematical Techniquesused in FEM analysis and use of them in Structural and thermal problem

TEXT BOOK:

1. Reddy. J.N., "An Introduction to the Finite Element Method", 3rd Edition, Tata McGraw-Hill, 2005

2. Seshu, P, “Text Book of Finite Element Analysis”, Prentice-Hall of India Pvt. Ltd., New Delhi, 2007.REFERENCE BOOKS:
1. Rao, S.S., “The Finite Element Method in Engineering”, 3rd Edition, Butterworth Heinemann, 2004
2. Logan, D.L., “A first course in Finite Element Method”, Thomson Asia Pvt. Ltd., 2002
3. Robert D. Cook, David S. Malkus, Michael E. Plesha, Robert J. Witt, “Concepts and Applications ofFinite Element Analysis”, 4th Edition, Wiley Student Edition, 2002.
4. Chandrupatla & Belagundu, “Introduction to Finite Elements in Engineering”, 3rd Edition, PrenticeHall College Div, 1990
5. Bhatti Asghar M, “Fundamental Finite Element Analysis and Applications”, John Wiley & Sons, 2005(Indian Reprint 2013)

UNIT I

FINITE ELEMENT FORMULATION OF BOUNDARY VALUE PROBLEMS

1.1 INTRODUCTION

The finite element method (FEM), or finite element analysis (FEA), is based on the idea of building a complicated object with simple blocks

Examples:
Lego (kids 'play) BuildingsApproximation of the area of a circle:

Why Finite Element Method?

- Design analysis: hand calculations, experiments, and computer simulations
- FEM/FEA is the most widelyapplied computer simulation method in engineering
- Closely integrated with CAD/CAM applications

1.1.1 A Brief History of the FEM

- 1943 --- Courant (variational method)
- 1956 --- Turner, clough, martin and top(stiffness)
- 1960 --- Clough (finite element plan problems)
- 1970 --- Applications on mainframe computer
- 1980 --- Microcomputers, pre and post processors
- 1990 --- Analysis of large structural systems

1.1.2 General Methods of the Finite Element Analysis

1. Force Method – Internal forces are considered as the unknowns of the problem.
2. Displacement or stiffness method – Displacements of the nodes are considered as theunknowns of the problem.

1.1.3 General Steps of the Finite Element Analysis

- Discretization of structure
- Numbering of Nodes and Elements
- Selection of Displacement function or interpolation function
- Define the material behavior by using Strain – Displacement and Stress – Strain relationships

- Derivation of element stiffness matrix and equations
- Assemble the element equations to obtain the global or total equations
- Applying boundary conditions
- Solution for the unknown displacements computation of the element
- strains and stresses from the nodal displacements
- Interpret the results (post processing).

1.1.4 Objectives of This FEM

- Understand the fundamental ideas of the FEM
- Know the behavior and usage of each type of elements coveredin this course
- Be able to prepare a suitable FE model for given problems
- Can interpret and evaluate the quality of the results (know the physics of the problems)
- Be aware of the limitations of the FEM (don't misuse the
- FEM - a numerical tool)

1.1.5 Applications of FEM in Engineering

- Mechanical/Aerospace/Civil/Automobile
  Engineering Structure analysis
  (static/dynamic, linear/nonlinear)
  Thermal/fluid flows
- Electromagnetics
- Geomechanics
- Biomechanics

1.2 WEIGHTED RESIDUAL METHOD

It is a powerful approximate procedure applicable to several problems. For non – structural problems, the method of weighted residuals becomes very useful. It has many types. The popular four methods are,

1. Point collocation method,

Residuals are set to zero at n different locations X_i, and the weighting function w_i is denoted as $\delta(x - x_i)$.

$$\int \delta(x - x_i)\, R(x; a_1, a_2, a_3 \ldots a_n)\, dx = 0$$

2. Subdomain collocation method

3. Least square method,

$\int [R(x; a_1, a_2, a_3 \ldots a_n)]^2 dx$ = minimum.

4. Galerkin's method. $w_i = N_i(x)$

$\int N_i(x) [R(x; a_1, a_2, a_3 \ldots a_n)]^2 dx = 0, \qquad i = 1, 2, 3, \ldots n.$

Problem I

Find the solution for the following differential equation.

$$EI \frac{d4u}{dx} - q_o = 0$$

The boundary conditions are $u(0)=0$, $\frac{du}{dx}(0)=0$,

$\frac{d2u}{dx2}(L)=0$, $\frac{d3u}{dx3}(L)=0$,

Given: The governing differential equation

$$EI \frac{d4u}{dx} - q_o = 0$$

Solution: assume a trial function

Let $u(x) = a_0+a_1x+a_2x^2+a_3x^3+a_4x^4.....$

Apply 1st boundary condition

$x=0, \quad u(x)=0$

$0=a_0+0$

$a_0=0$

Apply 2nd boundary condition

$x=0, \quad \frac{du}{dx}=0$

$a_1=0$

Apply 3rd bounaary condition

$x=L, \quad \frac{d2u}{dx2}=0$

$a_2=-[3a_3L+6a_4L^2]$

Apply 4th boundary condition

$x=L, \quad \frac{d3u}{dx3}=0$

$a_3=-4a_4L$

Substitute a_0, a1, a2 and $a_3$values in trial function

$u(x)= 0+0-[3a_3L+6a_4L^2] -4a_4L$

$u(x)= a_4[6 L^2x^2-4 Lx^3+ x^4]$

$\frac{du}{dx}= a_4[6 L^2 (2x)-12 Lx^2+ 4x^3]$

$\frac{d4u}{dx4}=24 a_4$

$R= EI \frac{d4u}{dx} - q_o = 0$

$a_4=\frac{qo}{24EI}$

Substitute a values in u(x)

$u(x) = \frac{qo}{24EI}[x_4-4Lx_3+6L^2x_2]$

Result:

Final solution $u(x) = \frac{qo}{24EI}[x_4-4Lx_3+6L^2x_2]$

Problem 2

The differential equation of a physical phenomenon is given by

$$\frac{d^2y}{dx^2} + y = 4, \quad 0 \leq x \leq 1$$

The boundary conditions are: y(0)=0

y(1)=1

Obtain one term approximate solution by using galerkin method

Solution:

Here the boundary conditions are not homogeneous so we assume a trial function as,

$y=a_1x(x-1)+x$

first we have to verify whether the trial function satisfies the boundary condition or not

$y=a_1x(x-1)+x$

when x=0, y=0

x=1, y=1

Resulual R:

$Y=a_1x(x-1)+x=a_1(x^2-x)+x$

$\frac{dy}{dx}=a_1(2x-1)+1$

$\frac{d^2y}{dx^2} = 2a_1$

Substitute $\frac{d^2y}{dx^2}$ value in given differential equation.

$2a_1+y=4x$

Substitute y vlue

$R=2a_1+a_1x(x-1)+x-4x$

In galerkin's method

$\int_0^1 wi\, R\, dx$

Substitute w_i and R value in equation

$a_1=0.83$

So one of the approximate solution is, $y= 0.83x(x-1)+x$

$= 0.83x^2-0.83x+x$

$y=0.83\, x^2+0.17x$

Problem 3

Find the deflection at the center of a simply supported beam of span length l subjected to uniform distributed load throughout its length as shown using (a) point collection method (b) Sub-domain method (c)least squared and (d) galerkin'n method.

Solution:

$$\mathrm{EI}\frac{d4y}{dx4}-\omega = 0, \quad 0\leq x \leq l$$

The boundaay condition are y= 0x= 0and y= l

$\mathrm{EI}\frac{d4y}{dx4}$=0 at x=0 and x= l

Where, $\mathrm{EI}\frac{d4y}{dx4} = M$

Let us select the trail function for deflection as,

$y= a \sin \pi x/l$

1.3 THE GENERAL WEIGHTED RESIDUAL STATEMENT

After understanding the basic techniques and successfully solved a few problem general weighted residual statement can be written as

R dx=0 for i= 1,2,.….n

Where $w_i=N_i$

The better result will be obtained by considering more terms in polynomial and trigonometric series.

1.4 WEAK FORMULATION OF THE WEIGHTED RESIDUAL STATEMENT.

The analysis in Section as applied to the model problem provides an attractive perspective to the solution of certain partial differential equations: the solution is identified with a "point", which minimizes an appropriately constructed functional over an admis- sible function space. Weak (variational) forms can be made fully equivalent to respective strong forms, as evidenced in the discussion of the weighted residual methods, under certain smoothness assumptions. However, the equivalence between weak (variational) forms and variational principles is not guaranteed: indeed, there exists no general method of construct-

ing functionals I [u], whose extremization recovers a desired weak (variational) form. In thissense,

only certain partial differential equations are amenable to analysis and solution by variational methods.

Vainberg's theorem provides the necessary and sufficient condition for the equivalence of a weak (variational) form to a functional extremization problem. If such equivalence holds, the functional is referred to as a potential.

Theorem (Vainberg)
Consider a weak (variational) form

$G(u, \delta u) := B(u, \delta u) + (f, \delta u) + (\bar{q}, \delta u)_{\Gamma q} = 0$,

where $u \in U$, $\delta u \in U0$, and f and $\bar{q}$ are independent of u. Assume that G pos- sesses a Gˆateaux derivative in a neighborhood N of u, and the Gˆateaux differen- tial $D\delta u1\ B(u, \delta u2)$ is continuous iin u att evvery poinntt of N .

Then, the necessary and sufficient condition for the above weak form to be derivable from a potential in N is that

$$D\delta u1\ G(u, \delta u2) = D\delta u2\ G(u, \delta u1),$$

Namely that $D\delta u1\ G(u, \delta u2)$ be symmetric for all $\delta u1, \delta u2 = U0$ and all u = N .

Preliminary to proving the above theorem, introduce the following two lemmas:

Lemma 1 Show that $D_V I[u] = \lim$

In the above derivation, note that operations and $|\omega=0$ are not interchangeable (as they both refer to the same variable ω), while lim $\Delta\omega\to0$ and $|\omega=0$ are interchangeable, conditional upon sufficient smoothness of I [u].

Lemma 2 (Lagrange's formula)

Let I [u] be a functional with Gateaux derivatives everywhere, and u, u + δu be any points of U. Then,

$$I[u+\delta u]-I[u]=D\delta u\, I[u+\varrho\,\delta u] \qquad 0<\varrho<1.$$

To prove Lemma 2, fix u and u + δu in U, and define function f on R as

$$f(\omega):=I[u+\omega\,\delta u].$$

It follows that

$$F=\frac{df}{d\omega}=\lim_{\Delta\omega\to0}\frac{f(\omega+\Delta\omega)-f(\omega)}{\Delta\omega}$$

$$=\lim_{\Delta\omega\to0}\frac{I[u+\omega\,\delta u+\Delta\omega\,\delta u]-I[u+\omega\,\delta u]}{\Delta\omega}=D\delta u\, I[u+\omega\,\delta u],$$

Where Lemma 1 was invoked. Then, u s i n g the standard mean-value theorem of calculus,

1.5 PIECE WISE CONTINUOUS TRIAL FUNCTION

In weighted residual method the polynomial and trigonometric series are used as trial function. This trial function is a single composite function and it is valid over the entire solution domain this assumed trial function solution should match closely to the exact solution of the differential equation and the boundary conditions, it is nothing but a process of curve fitting. This curve fitting is carried out by piecewise method i.e., the more numbers of piece leads better curve fit. Piecewise method can be explained by the following simple problem.

We know that the straight line can be drawn through any two points.

Let, f(x)=sin— is the approximated function for straight line segments.

One straight line segment

Two straight line segment

One Spring Element

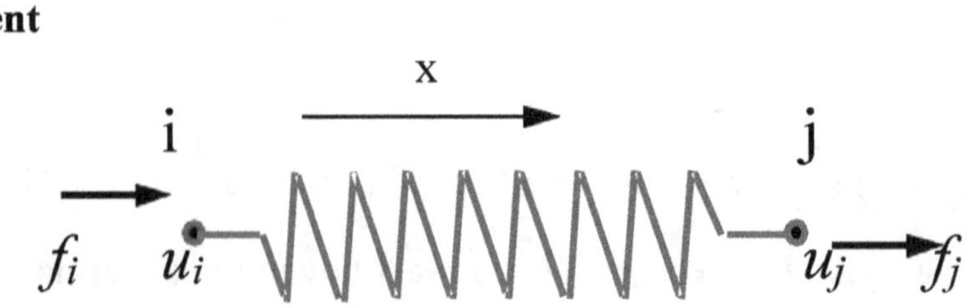

Two nodes: i, j
Nodal displacements: u_i, u_j (in, m, mm)
Nodal forces: f_i, f_j (lb, Newton) Spring constant (stiffness): k (lb/in, N/m, N/mm)

Spring force-displacement relationship:

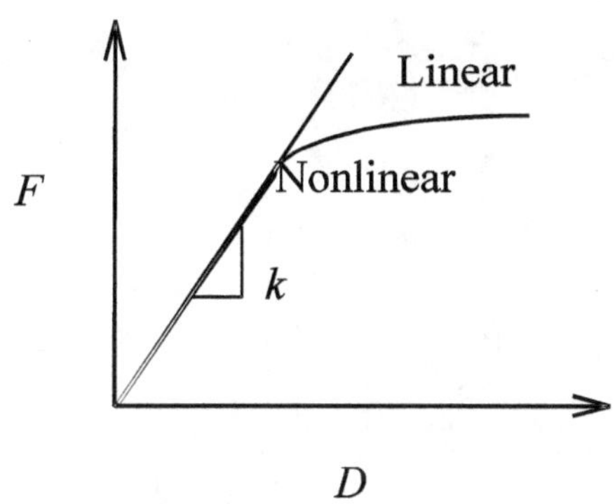

$k \quad F/ \quad$ (> 0) is the force needed to produce a unit stretch.

We only consider linear problems in this introductory course. Consider the equilibrium of forces for the spring.

At node 1 we have

$$f_i \qquad F \qquad k(u_j \qquad u_i) \qquad ku_i \qquad ku_j$$

and at node j,

$$f_j \qquad F \qquad k(u_j \qquad u_i) \qquad ku_i \qquad ku_j$$

In matrix form,

$$\begin{matrix} k & k & u_i & & f_i \\ k & k & u_j & & f_j \end{matrix}$$

or, where

(element) stiffness matrix

u = (element nodal) displacement vector

f = (element nodal) force vector

Note:

That k is symmetric. Is k singular or non singular? That is, can we solve the equation? If not, why?

Problem 4

To find the deformation of the shape

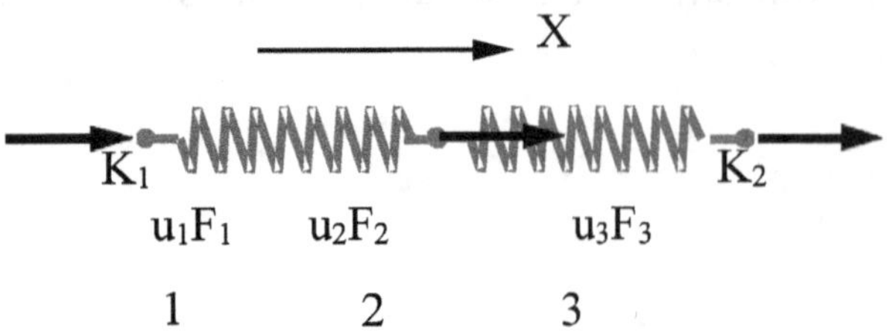

For element 1,

$$k_1 \quad k_1 \quad u_2 \quad f_1^2$$
$$k_1 \quad k_1 \quad u_3 \quad f_2^2$$

element 2,

$$k_2 \quad k_2 \quad u_2 \quad f^2$$
$$k_2 \quad k_2 \quad u_3 \quad f^2$$

where f_I at node 2 F_2

M is the (internal) force acting on local node *i* of element Consider the quilibrium of forces at node

Checking the Results

Deformed shape of the structure

Balance of the external forces

Order of magnitudes of the numbers

Notes about the Spring Elements

Suitable for stiffness analysis

Not suitable for stress analysis of the spring itself

Can have spring elements with stiffness in the lateral direction,

Spring elements for torsion, etc.

1.6 EXAMPLES OF A BAR FINITE ELEMENT

The finite element method can be used to solve a variety of problem types in engineering, mathematics and science. The three main areas are mechanics of materials, heat transfer and fluid mechanics. The one-dimensional spring element belongs to the area of mechanics of materials, since it deals with the displacements, deformations and stresses involved in a solid body subjected to external loading.

Element dimensionality:

An element can be one-dimensional, two-dimensional or three-dimensional. A spring element is classified as one-dimensional.

Geometric shape of the element

The geometric shape of element can be represented as a line, area, or volume. The one-dimensional spring element is defined geometrically as:

Spring law

The spring is assumed to be linear. Force (f) is directly proportional to deformation (Δ) via the spring constant k, i.e.

$$f = k\Delta u$$

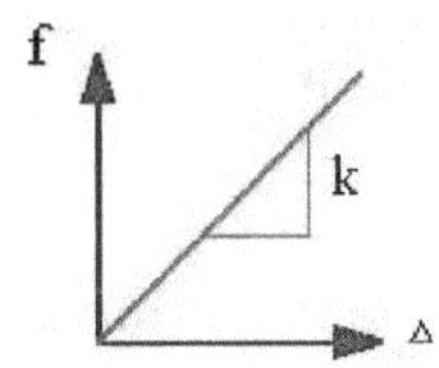

Types of degrees of freedom per node

Degrees of freedom are displacements and/or rotations that are associated with a node. A one-dimensional spring element has two translational degrees of freedom, which include, an axial(horizontal) displacement (u) at each node.

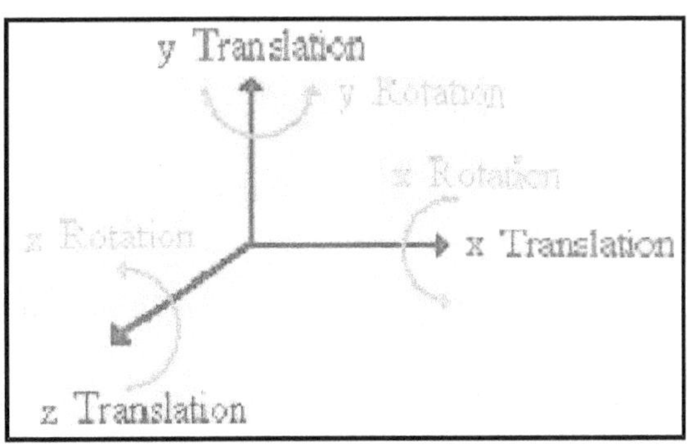

Element formulation

There are various ways to mathematically formulate an element. The simplest and limited approach is the direct method. More mathematically complex and general approaches are energy (variation) and weighted residual methods.

The direct method uses the fundamentals of equilibrium, compatibility and spring law from a sophomore level mechanics of material course. We will use the direct method to formulate the one-dimensional spring element because it is simple and based on a physical approach.

The direct method is an excellent setting for becoming familiar with such basis concepts of linear algebra, stiffness, degrees of freedom, etc., before using the mathematical formulation approaches as energy or weighted residuals.

Assumptions

Spring deformation

The spring law is a linear force-deformation as follows:

$f = k\,\Delta$

f - Spring Force (units: force)

k - Spring Constant (units: force/length)

Δ - Spring Deformation (units: length)

Spring Behaviour:

A spring behaves the same in tension and compression.

Spring Stiffness:

Spring stiffness k is always positive, i.e., k>0, for a physical linear system.

Nodal Force Direction:

Loading is uniaxial, i.e., the resultant force is along the element. Spring has no resistance to lateral force.

Weightless Member:

Element has no mass (weightless).

Node Location:

The geometric location of nodes I and J cannot coincide, i.e., $x_i \neq x_j$. The length of the element is only used to visually see the spring.

A column of K_E is a vector of nodal loads that must be applied to an element to sustain a deformed state in which responding nodal DOF has unit value and all other nodal DOF are zero. In other words, a column of K_E represents an equilibrium problem.

Slope continuity $\left(\frac{du}{dx}\right)$ not required at inter – element interface (node).

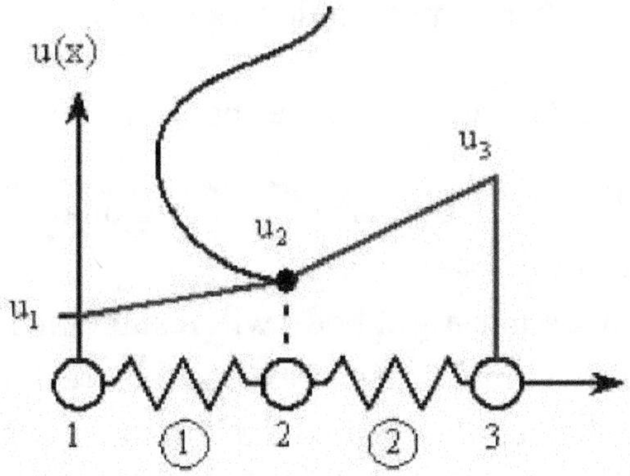

u_1 – Axial displacement of node 1.
u_2 – Axial displacement of node 2.
u_3 – Axial displacement of node 3.

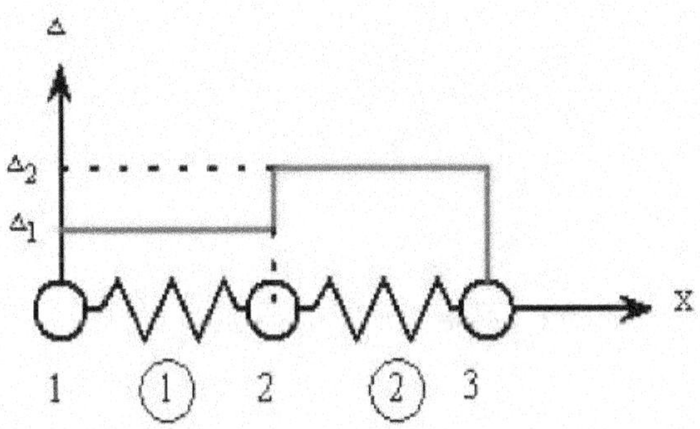

Δ_1 – Deformation of element ①.

Δ_2 – Deformation of element ②.

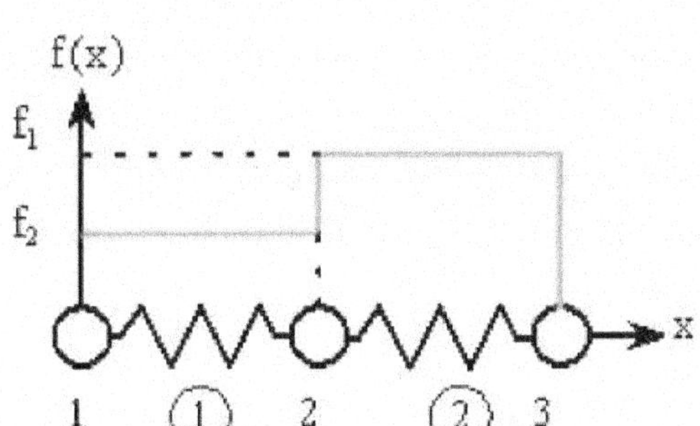

f_1 – Internal axial force of element ①.
f_2 – Internal axial force of element ②.

1.6.1 Rigid Body

A body is considered rigid if it does not deform when a force is applied. Consider rigid and non-rigid bars subjected to a gradually applied axial force of increasing magnitude as shown.

The reader should note the following characteristics of rigid and non-rigid (flexible) bodies:

- Force Magnitude - Even if forces are large, a rigid body does not deform. A non-rigid body will deform even if a force is small. In reality, all bodies deform.
- Failure - A rigid body does not fail under any load; while a non-rigid body will result either in ductile or brittle failure when the applied load causes the normal stress to exceed the breaking (fracture) stress σ_b of the material. Brittle failure occurs when the applied load on the non-rigid bar shown above causes the breaking strength of the bar to be exceeded.
- Material - The material is not considered in a rigid body. Since a rigid body does not deform (ε = 0) this is equivalent to an infinite modulus of elasticity. In contrast the modulus of elasticity for a non-rigid material is finite, e.g., for steel, $E_{steel} = 30 \times 10^6$ psi. (200 GPa). For rigid and non-rigid bars the material laws are:

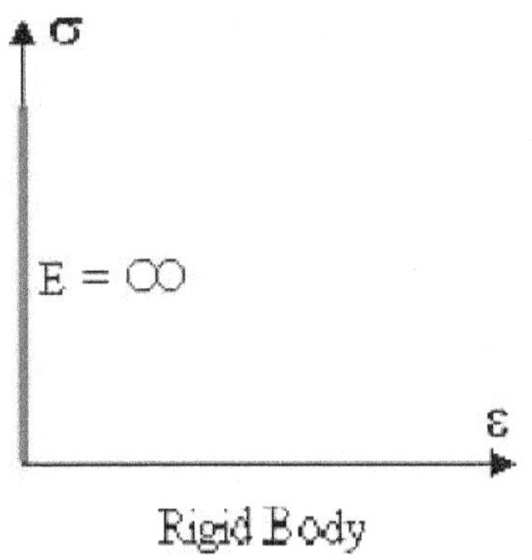

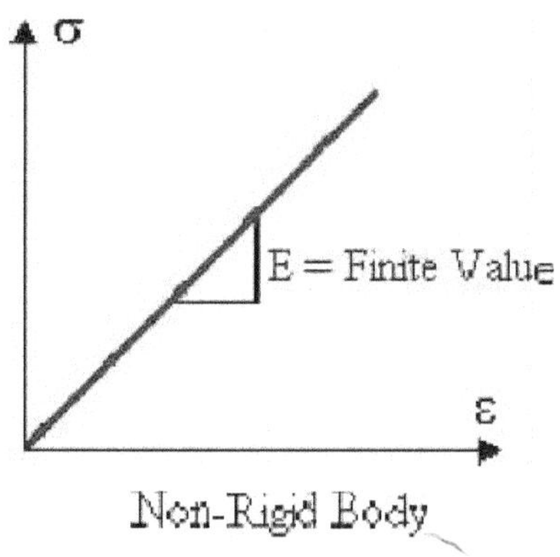

Rigid Body Motion

Rigid body motion occurs when forces and/or moments are applied to an unrestrained mesh (body), resulting in motion that occurs without any deformations in the entire mesh (body). Since no strains (deformations) occur during rigid body motion, there can be no stresses developed in the mesh.

A rigid body in general can be subjected to three types of motion, which are translation, rotation about a fixed axis, and general motion which consists of a combination of both translation and rotation. These three motion types are as follows:

Translation - If any line segment on the body remains parallel to its original direction during the motion, it is said to be in translation. When the path of motion is along a straight line, the motion is called rectilinear translation, while a curved path is considered as a curvilinear translation. The curvilinear motion shown below is a combination of two translational motions, one horizontal motion and one vertical motion.

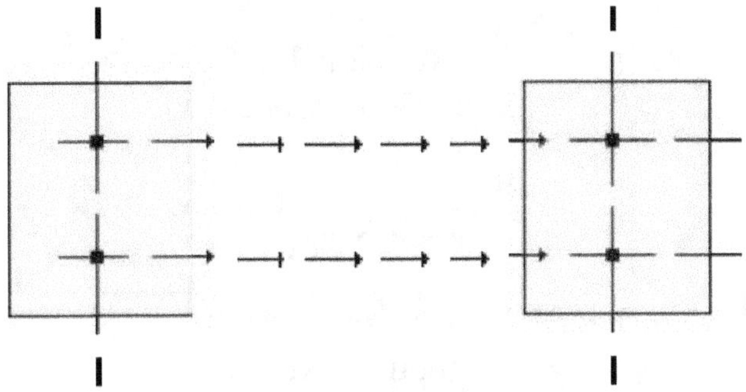

Rectilinear Translation

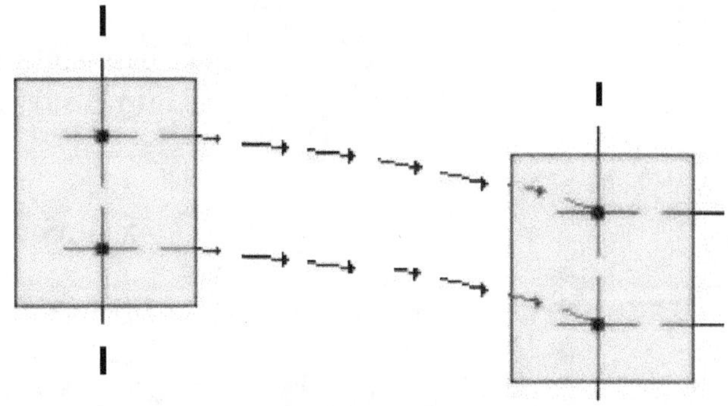

Curvilinear Translation

Rotation About a Fixed Axis - If all the particles of a rigid body move along circular paths, except the ones which lie on the axis of rotation, it is said to be in rotation about a fixed **a**xis.

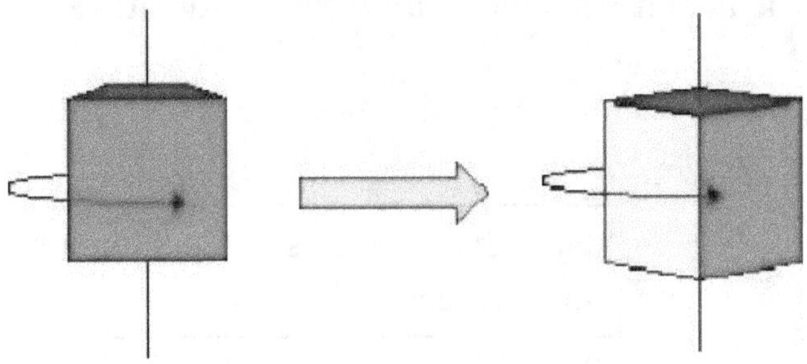

Rotation about a Fixed Axis

General Motion - Any motion of a rigid body that consists of the combination of both translations

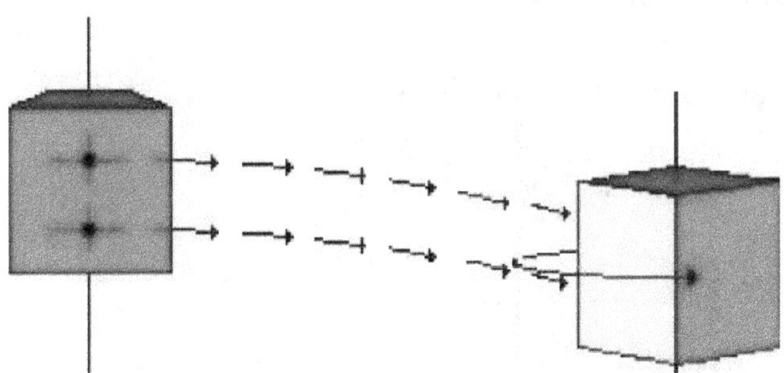

General Motion

There are six rigid body modes in general three-dimensional situation; three translational along the x, y, and z axes and three rotational about x, y, and z axes. Illustrations of these rigid body modes are presented as follows:

Translational Rigid Body Modes	Rotational Rigid Body Modes
x-direction	about x-axis
y-direction	about y-axis
z-direction	about z-axis

1-D 3-NODED QUADRATIC BAR ELEMENT

Problem 6

A single 1-D 3-noded quadratic bar element has 3 nodes with local coordinates as shown in Figure

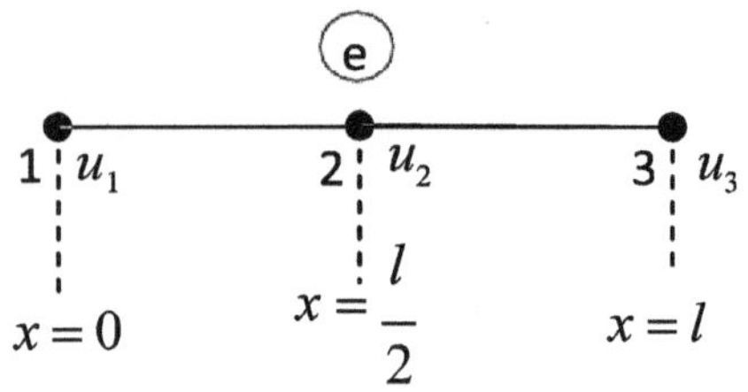

Note that node 2 is at the midpoint of the element.

The chosen approximation function for the field variable u is $u = a + bx + cx^2$

Let the field variable u have values u_1, u_2 and u_3 at nodes 1, 2 and 3, respectively.

To find the unknowns *a*, *b* and *c*, we apply the boundary conditions

at $x = 0$, $\quad u = u_1 \quad \Rightarrow \quad u_1 = a \quad \Rightarrow \quad a = u_1$

at $x = \frac{l}{2}$, $\quad u = u_2 \quad \Rightarrow \quad u_2 = a + b\frac{l}{2} + c\frac{l^2}{4}$

at $x = l$, $\quad u = u_3 \quad \Rightarrow \quad u_3 = a + bl + cl^2$

solving

$$b = \frac{-u_3 + 4u_2 - 3u_1}{l}$$

$$c = \left(\frac{2}{l^2}\right)(u_3 - 2u_2 + u_1)$$

Substituting the values of *a*, *b* and *c* in equation (1) and collecting the coefficients of u_1, u_2 and u_3

$$u = N_1u_1 + N_2u_2 + N_3u_3$$

Where

$$N_1 = 1 - 3\frac{x}{l} + 2\frac{x^2}{l^2} = \left(1-\frac{x}{l}\right)\left(1-\frac{2x}{l}\right) \qquad \frac{\partial N_1}{\partial x} = -\frac{3}{l} + 4\frac{x}{l^2}$$

$$N_2 = 4\frac{x}{l} - 4\frac{x^2}{l^2} = 4\frac{x}{l}\left(1-\frac{x}{l}\right) \qquad \frac{\partial N_2}{\partial x} = \frac{4}{l} - 8\frac{x}{l^2}$$

$$N_3 = -\frac{x}{l} + 2\frac{x^2}{l^2} = -\frac{x}{l}\left(1-\frac{2x}{l}\right) \qquad \frac{\partial N_3}{\partial x} = -\frac{1}{l} + 4\frac{x}{l^2}$$

Derivation of stiffness matrix for 1-D 3-noded quadratic bar element:

$$[B]^T = \begin{Bmatrix} -\frac{3}{l} + 4\frac{x}{l^2} \\ \frac{4}{l} - 8\frac{x}{l^2} \\ -\frac{1}{l} + 4\frac{x}{l^2} \end{Bmatrix}$$

$[D] = E$ for a bar element (1-D case - only axial stress (σ_x) and strain (ε_x) exist $\Rightarrow \sigma_x = E\,\varepsilon_x$)

$$\int_{volume} dV = \int_0^l A\,dx = A\int_0^l dx$$ since the cross-sectional area A is constant for the total length of the bar.

$$[k] = A\int_0^l \begin{Bmatrix} -\frac{3}{l} + 4\frac{x}{l^2} \\ \frac{4}{l} - 8\frac{x}{l^2} \\ -\frac{1}{l} + 4\frac{x}{l^2} \end{Bmatrix} E \left\langle \left(-\frac{3}{l} + 4\frac{x}{l^2}\right) \left(\frac{4}{l} - 8\frac{x}{l^2}\right) \left(-\frac{1}{l} + 4\frac{x}{l^2}\right) \right\rangle dx$$

$$[k] = AE\int_0^l \begin{bmatrix} \left(-\frac{3}{l}+4\frac{x}{l^2}\right)\left(-\frac{3}{l}+4\frac{x}{l^2}\right) & \left(-\frac{3}{l}+4\frac{x}{l^2}\right)\left(\frac{4}{l}-8\frac{x}{l^2}\right) & \left(-\frac{3}{l}+4\frac{x}{l^2}\right)\left(-\frac{1}{l}+4\frac{x}{l^2}\right) \\ \left(\frac{4}{l}-8\frac{x}{l^2}\right)\left(-\frac{3}{l}+4\frac{x}{l^2}\right) & \left(\frac{4}{l}-8\frac{x}{l^2}\right)\left(\frac{4}{l}-8\frac{x}{l^2}\right) & \left(\frac{4}{l}-8\frac{x}{l^2}\right)\left(-\frac{1}{l}+4\frac{x}{l^2}\right) \\ \left(-\frac{1}{l}+4\frac{x}{l^2}\right)\left(-\frac{3}{l}+4\frac{x}{l^2}\right) & \left(-\frac{1}{l}+4\frac{x}{l^2}\right)\left(\frac{4}{l}-8\frac{x}{l^2}\right) & \left(-\frac{1}{l}+4\frac{x}{l^2}\right)\left(-\frac{1}{l}+4\frac{x}{l^2}\right) \end{bmatrix} dx$$

$$[k] = \int_{Volume} [B]^T [D][B]\, dV$$

$$[B] = \left\langle \frac{\partial N_1}{\partial x} \quad \frac{\partial N_2}{\partial x} \quad \frac{\partial N_3}{\partial x} \right\rangle = \left\langle \left(-\frac{3}{l} + 4\frac{x}{l^2}\right) \left(\frac{4}{l} - 8\frac{x}{l^2}\right) \left(-\frac{1}{l} + 4\frac{x}{l^2}\right) \right\rangle$$

To determine K_{11} :

$$K_{11} = AE\int_0^l \left(\frac{-3}{l}+\frac{4x}{l^2}\right)\left(\frac{-3}{l}+\frac{4x}{l^2}\right)dx = AE\int_0^l \left(\frac{9}{l^2}-\frac{12x}{l^3}-\frac{12x}{l^3}+\frac{16x^2}{l^4}\right)dx$$

Integrating and applying limit we get,

$$K_{11} = AE\int_0^l \left(\frac{9}{l^2}-\frac{24x}{l^3}+\frac{16x^2}{l^4}\right)dx = AE\left[\frac{9x}{l^2}-\frac{24x^2}{2l^3}+\frac{16x^3}{3l^4}\right]_0^l = AE\left[\frac{9l}{l^2}-\frac{24l^2}{2l^3}+\frac{16l^3}{3l^4}\right]$$

$$K_{11} = AE\left[\frac{9}{l}-\frac{12}{l}+\frac{16}{3l}\right] = AE\left[\frac{27-36+16}{3l}\right] = \frac{AE}{3l}[7]$$

$$K_{11} = \frac{7AE}{3l}$$

To determine K_{12} (and **K_{21}**):

$$K_{12} = AE\int_0^l \left(-\frac{3}{l}+\frac{4x}{l^2}\right)\left(\frac{4}{l}-\frac{8x}{l^2}\right)dx = AE\int_0^l \left(-\frac{12}{l^2}+\frac{24x}{l^3}+\frac{16x}{l^3}-\frac{32x^2}{l^4}\right)dx$$

$$K_{12} = AE\left[-\frac{12x}{l^2}+\frac{40x^2}{}-\frac{32x^3}{3l^4}\right]_0^l$$

$$K_{12} = AE\left[-\frac{12x}{l^2}+\frac{40x^2}{2l^3}-\frac{32x^3}{3l^4}\right]_0^l = AE\left[-\frac{12l}{l^2}+\frac{40l^2}{2l^3}-\frac{32l^3}{3l^4}\right] = AE\left[-\frac{12}{l}+\frac{20}{l}-\frac{32}{3l}\right] = AE\left[\frac{-36+60-32}{3l}\right]$$

$$K_{12} = \frac{AE}{3l}[-8] = K_{21}$$

To determine K_{13} (and **K_{31}**):

$$K_{13} = AE\int_0^l \left(-\frac{3}{l}+\frac{4x}{l^2}\right)\left(-\frac{1}{l}+\frac{4x}{l^2}\right)dx = AE\int_0^l \left(\frac{3}{l^2}-\frac{12x}{l^3}-\frac{4x}{l^3}+\frac{16x^2}{l^4}\right)dx$$

$$K_{13} = AE\int_0^l \left(\frac{3}{l^2}-\frac{16x}{l^3}+\frac{16x^2}{l^4}\right)dx = AE\left[\frac{3x}{l^2}-\frac{16x^2}{2l^3}+\frac{16x^3}{3l^4}\right]_0^l = AE\left[\frac{3l}{l^2}-\frac{16l^2}{2l^3}+\frac{16l^3}{3l^4}\right] = AE\left[\frac{3}{L}-\frac{8}{L}+\frac{16}{3L}\right]$$

$$K_{13} = AE\left[\frac{9-24+16}{3l}\right] = \frac{AE}{3l}[1] = K_{31}$$

To determine K_{22}

$$K_{22} = AE\int_0^l \left(\frac{4}{l} - \frac{8x}{l^2}\right)\left(\frac{4}{l} - \frac{8x}{l^2}\right)dx = AE\int_0^l \left(+\frac{16}{l^2} - \frac{32x}{l^3} - \frac{32x}{l^3} + \frac{64x^2}{l^4}\right)dx$$

$$K_{22} = AE\int_0^l \left(\frac{16}{l^2} - \frac{64x}{l^3} + \frac{64x^2}{l^4}\right)dx = AE\left[\frac{16x}{l^2} - \frac{64x^2}{} + \frac{64x^3}{3l^4}\right]_0^l$$

$$K_{22} = AE\left[\frac{16l}{l^2} - \frac{64l^2}{2l^3} + \frac{64l^3}{3l^4}\right] = AE\left[\frac{16}{l} - \frac{32}{l} + \frac{64}{3l}\right] = AE\left[\frac{48-96+64}{3l}\right] = \frac{AE}{3l}[16]$$

$$K_{22} = \frac{16\,AE}{3l}$$

To determine K_{23} (and K_{32})

$$K_{23} = AE\int_0^l \left(-\frac{4}{l^2} + \frac{24x}{l^3} - \frac{32x^2}{l^4}\right)dx = AE\left[-\frac{4x}{l^2} + \frac{24x^2}{2l^3} - \frac{32x^3}{3l^4}\right]_0^l$$

$$K_{23} = AE\left[-\frac{4l}{l^2} + \frac{24l^2}{2l^3} - \frac{32l^3}{3l^4}\right] = AE\left[\frac{-4}{l} + \frac{12}{l} - \frac{32}{3l}\right] = AE\left[\frac{-12+36-32}{3l}\right] = \frac{AE}{3l}[-8]$$

$$K_{23} = -\frac{8AE}{3l} = K_{32}$$

To determine K_{33}

$$K_{33} = AE\int_0^l \left(-\frac{1}{l} + \frac{4x}{l^2}\right)\left(-\frac{1}{l} + \frac{4x}{l^2}\right)dx = AE\int_0^l \left(\frac{1}{l^2} - \frac{4x}{l^3} - \frac{4x}{l^3} + \frac{16x^2}{l^4}\right)dx$$

$$K_{23} = AE\int_0^l \left(\frac{4}{l} - \frac{8x}{l^2}\right)\left(-\frac{1}{l} + \frac{4x}{l^2}\right)dx = AE\int_0^l \left(-\frac{4}{l^2} + \frac{16x}{l^3} + \frac{8x}{l^3} - \frac{32x^2}{l^4}\right)dx$$

$$K_{33} = AE\int_0^l \left(\frac{1}{l^2} - \frac{8x}{l^3} + \frac{16x^2}{l^4}\right)dx = AE\left[\frac{x}{l^2} - \frac{8x^2}{2l^3} + \frac{16x^3}{3l^4}\right]_0^l$$

$$K_{33} = AE\left[\frac{l}{l^2} - \frac{8l^2}{2l^3} + \frac{16l^3}{3l^4}\right] = AE\left[\frac{1}{l} - \frac{4}{l} + \frac{16}{3l}\right] = AE\left[\frac{3-12+16}{3l}\right] = \frac{AE}{3l}[7]$$

$$K_{33} = \frac{7AE}{3l}$$

Assembling, we get $[k] = \frac{AE}{3L}\begin{bmatrix} 7 & -8 & 1 \\ -8 & 16 & -8 \\ 1 & -8 & 7 \end{bmatrix}$

1.7 PRINCIPLE OF STATIONERY TOTAL POTENTIAL (PSTP)

1.7.1 Potential energy in elastic bodies

Potential energy is the capacity to do the work by the force acting on deformable bodies; the forces acting on a body may be classified as external forces and internal forces. External forces are the applied loads while internal force is the stresses developed in the body. Hence the total potential energy is the sum of internal and external potential energy.

Consider a spring mass system let its stiffness be k and length L, due to a force P let it extend by u

The load P moves down by distance u. hence it loses its capacity to do work by P u. the external potential energy in this case is given by.

$$H = -P\,u$$

Average force = $\frac{K\,u}{2}$

The energy stored in the spring due to strain = Average force x Deflection

$$= \frac{K\,u}{2} \times u$$

$$= \frac{1}{2} K u^2$$

Total potential energy in the spring $\pi = \frac{1}{2} K u^2 - P u$

1.7.2 Principle of Minimum Potential Energy

From the expression for total potential energy,

$$\pi = U + H$$

$$\delta\pi = \delta U + \delta H$$

In principle of virtual work $\delta U = \delta H$

$$\delta\pi = 0$$

Hence we can conclude that a deformable body is in equilibrium when the potential energy is having stationary value.

Hence the principle of minimum potential energy states among all the displacement equations that internal compatibility and the boundary condition those that also satisfy the equation of equilibrium make the potential energy a minimum is a stable system

Problem 7

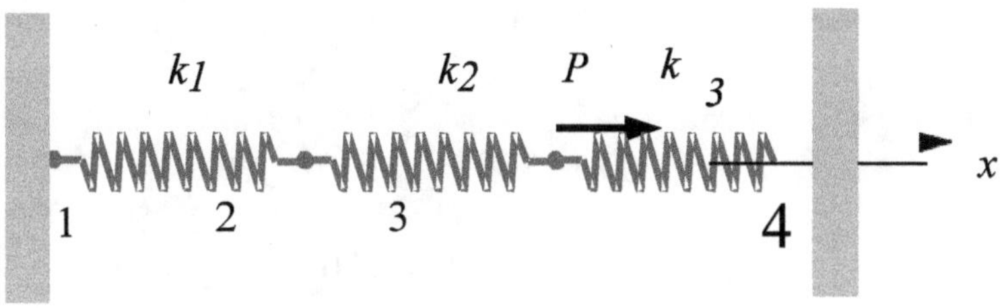

Given: For the spring system shown above,

k_1 100 N / mm,

k_2 200 N / mm,

k_3 100 N / mm

P 500 N,

u_1 0

u_4 0

Find:

(a) The global stiffness matrix

(b) Displacements of nodes 2 and 3

(c) The reaction forces at nodes 1 and 4

(d) the force in the spring 2

Solution:

(a) The element stiffness matrices are

$$\mathrm{k}_1 \quad \begin{matrix} 100 & 100 \\ 100 & 100 \end{matrix} \quad \text{(N/mm)} \qquad (1)$$

$$\mathrm{k}_2 \quad \begin{matrix} 200 & 200 \\ 200 & 200 \end{matrix} \quad \text{(N/mm)} \qquad (2)$$

$$\mathrm{k}_3 \quad \begin{matrix} 100 & 100 \\ 100 & 100 \end{matrix} \quad \text{(N/mm)} \qquad (3)$$

$$
\mathbf{K} \quad
\begin{array}{ccccc}
u_1 & u_2 & u_3 & u_4 & \\
100 & 100 & 0 & 0 & \\
100 & 100 & 200 & 200 & 0 \\
0 & 200 & 200 & 100 & 100 \\
 & 200 & 300 & 100 & \\
 & 0 & 100 & 100 &
\end{array}
$$

which is *symmetric* and ***banded***.

Equilibrium (FE) equation for the whole system is

$$
\begin{array}{cccc}
100 & 100 & 0 & 0 \\
100 & 300 & 200 & 0 \\
0 & 200 & 300 & 100 \\
0 & 0 & 100 & 100
\end{array}
\quad
\begin{array}{c}
u_1 \\ u_2 \\ u_3 \\ u_4
\end{array}
\quad
\begin{array}{c}
F_1 \\ 0 \\ P \\ F_4
\end{array}
\qquad (4)
$$

(b) Applying the BC

$$
\begin{array}{cc}
300 & 200 \\
200 & 300
\end{array}
\quad
\begin{array}{c}
u_2 \\ u_3
\end{array}
\quad
\begin{array}{c}
0 \\ P
\end{array}
\qquad (5)
$$

Solving Eq.(5), we obtain

$$
\begin{array}{ccc}
u_2 & P/250 & 2 \\
u_3 & 3P/500 & 3
\end{array}
\quad \text{(mm)}
\qquad (6)
$$

(c) From the 1st and 4th equations in (4), we get the reaction forces

$$
\begin{array}{ccc}
F_1 & 100u_2 & 200 \text{ (N)} \\
F_4 & 100u_3 & 300 \text{ (N)}
\end{array}
$$

(d) The FE equation for spring (element) 2 is

$$\begin{matrix} 200 & 200 & u_i \\ 200 & 200 & u_j \end{matrix} \qquad \begin{matrix} f_i \\ f_j \end{matrix}$$

Here $i = 2, j = 3$ for element 2. Thus we can calculate the spring force as

Problem 8

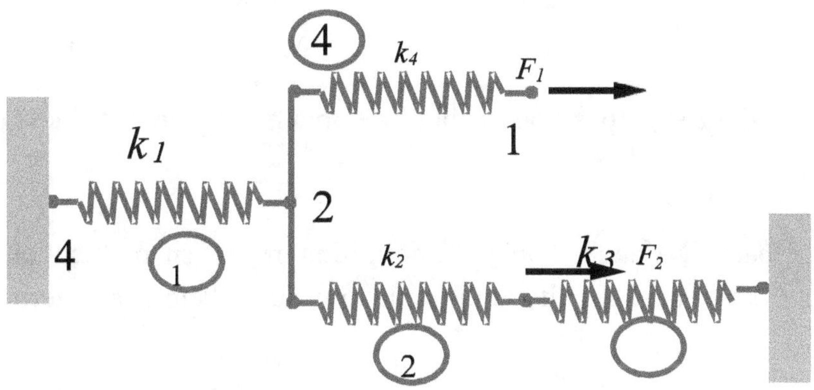

For the spring system with arbitrarily numbered nodes and elements, as shown above, find the global stiffness matrix.

Solution:

First we construct the following

Element Connectivity Table

Element	*Node i (1)*	*Node j (2)*
1	4	2
2	2	3
3	3	5
4	2	1

Which specifies the global node numbers corresponding to the local node numbers for each element? Then we can write the element stiffness matrices as follows

$$\mathbf{k} \quad \begin{matrix} u_4 & u_2 \\ k_1 & k_1 \\ k_1 & k_1 \end{matrix} \qquad \mathbf{k}_2 \quad \begin{matrix} u_2 & u_3 \\ k_2 & k_2 \\ k_2 & k_2 \end{matrix}$$

$$\mathbf{k}_3 \quad \begin{matrix} u_3 & u_5 \\ k_3 & k_3 \\ k_3 & k_3 \end{matrix} \qquad \mathbf{k}_4 \quad \begin{matrix} u_2 & u_1 \\ k_4 & k_4 \\ k_4 & k_4 \end{matrix}$$

Finally, applying the superposition method, we obtain the global stiffness matrix as follows

We may note that N1 and N2 obey the definition of shape function that is the shape function will have a value equal to unity at the node to which it belong and zero value at other nodes.

$$\mathbf{K} \quad \begin{matrix} u_1 & u_2 & u_3 & u_4 & u_5 \\ k_4 & k_4 & 0 & 0 & 0 \\ k_4 & k_1 \; k_2 \; k_4 & k_2 & k_1 & 0 \\ 0 & k_2 & k_2 \; k_3 & 0 & k_3 \\ 0 & k_1 & 0 & k_1 & 0 \\ 0 & 0 & k_3 & 0 & k_3 \end{matrix}$$

1.8 RAYLEIGH – RITZ METHOD (VARIATIONAL APPROACH)

It is useful for solving complex structural problems. This method is possible only if a suitable functional is available. Otherwise, Galerkin's method of weighted residual is used.

Problems (I set)

1. A simply supported beam subjected to uniformly distributed load over entire span. Determine the bending moment and deflection at midspan by using Rayleigh – Ritz method and compare with exact solutions.

2. A bar of uniform cross section is clamed at one end and left free at another end and it is subjected to a uniform axial load P. Calculate the displacement and stress in a bar by using two terms polynomial and three terms polynomial. Compare with exact solutions.

1.9 ADVANTAGES OF FINITE ELEMENT METHOD

1. FEM can handle irregular geometry in a convenient manner.
2. Handles general load conditions without difficulty
3. Non – homogeneous materials can be handled easily.
4. Higher order elements may be implemented.

1.10 DISADVANTAGES OF FINITE ELEMENT METHOD

1. It requires a digital computer and fairly extensive
2. It requires longer execution time compared with FEM.
3. Output result will vary considerably.

UNIT II

ONE DIMENSIONAL FINITE ELEMENT ANALYSIS

2.1 ONE DIMENSIONAL ELEMENTS

Bar and beam elements are considered as One Dimensional elements. These elements are often used to model trusses and frame structures.

- **Bar, Beam and Truss**

Bar is a member which resists only axial loads. A beam can resist axial, lateral and twisting loads. A truss is an assemblage of bars with pin joints and a frame is an assemblage of beam elements.

- **Stress, Strain and Displacement**

Stress is denoted in the form of vector by the variable x as σ_x, Strain is denoted in the form of vector by the variable x as e_x, Displacement is denoted in the form of vector by the variable x as u_x.

- **Types of Loading**

(1) Body force (f)

It is a distributed force acting on every elemental volume of the body. Unit is Force / Unit volume. Ex: Self weight due to gravity.

(2) Traction (T)

It is a distributed force acting on the surface of the body. Unit is Force / Unit area. But for one dimensional problem, unit is Force / Unit length. Ex: Frictional resistance, viscous drag and Surface shear.

(3) Point load (P)

It is a force acting at a particular point which causes displacement.

- **Finite Element Modeling**

It has two processes.

(1) Discretization of structure

(2) Numbering of nodes.

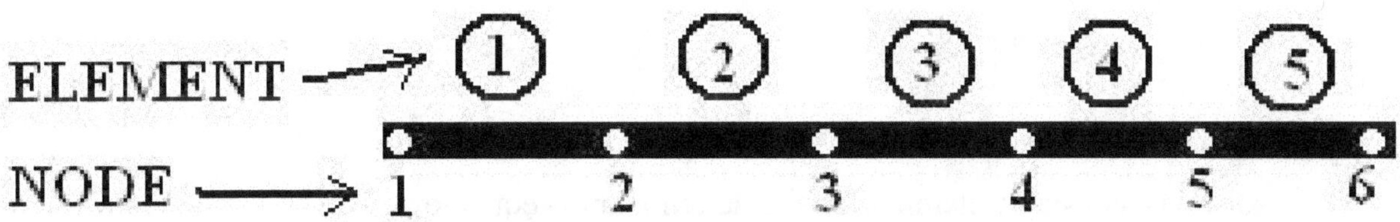

- **CO – ORDINATES**

  (A) Global co – ordinates,

  (B) Local co – ordinates and

  (C) Natural co –ordinates.

- **Natural Co – Ordinate (ε)**

  Integration of polynomial terms in natural co – ordinates for two dimensional elements can be performed by using the formula,

  α β γ

- **Shape function**

  $N_1N_2N_3$ are usually denoted as shape function. In one dimensional problem, the displacement

  $$u = \Sigma N_i u_i = N_1 u_1$$

  For two noded bar element, the displacement at any point within the element is given by,

  $$u = \Sigma N_i u_i = N_1 u_1 + N_2 u_2$$

  For three noded triangular element, the displacement at any point within the element is given by,

  $$u = \Sigma N_i u_i = N_1 u_1 + N_2 u_2 + N_3 u_3$$

  $$v = \Sigma N_i v_i = N_1 v_1 + N_2 v_2 + N_3 v_3$$

  Shape function need to satisfy the following

  (a) First derivatives should be finite within an element; (b) Displacement should be continuous across the element boundary

- **Polynomial Shape function**

  Polynomials are used as shape function due to the following reasons, (1) Differentiation and integration of polynomials are quite easy.

  (2) It is easy to formulate and computerize the finite element equations.

  (3) The accuracy of the results can be improved by increasing the order of

- **Properties of Stiffness Matrix**

  1. It is a symmetric matrix,
  2. The sum of elements in any column must be equal to zero,
  3. It is an unstable element. So the determinant is equal to zero.

- **Problem (I set)**

  1. A two noded truss element is shown in figure. The nodal displacements are $u_1 = 5$ mm and $u_2 = 8$ mm. Calculate the displacement at x = ¼, 1/3 and ½.

- **Problem (II set)**

  1. Consider a three bar truss as shown in figure. It is given that $E = 2 \times 10^5$ N/mm^2. Calculate
  (a) Nodal displacement,
  (b) Stress in each member and
  (c) Reactions at the support. Take Area of element 1 = 2000 mm^2, Area of element 2 = 2500 mm^2, Area of element 3 = 2500 mm^2.

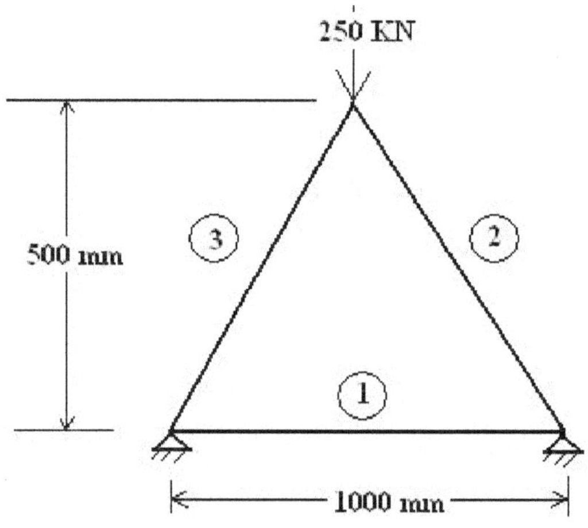

- **Types of beam**

  1. Cantilever beam,
  2. Simply Supported beam,
  3. Over hanging beam,
  4. Fixed beam and
  5. Continuous beam.

- **Types of Transverse Load**

  1. Point or Concentrated Load,
  2. Uniformly Distributed Load and
  3. Uniformly

➢ **Problem (III set)**

1. A fixed beam of length 2L m carries a uniformly distributed load of w (N/m) which runs over a length of L m from the fixed end. Calculate the rotation at Point B.

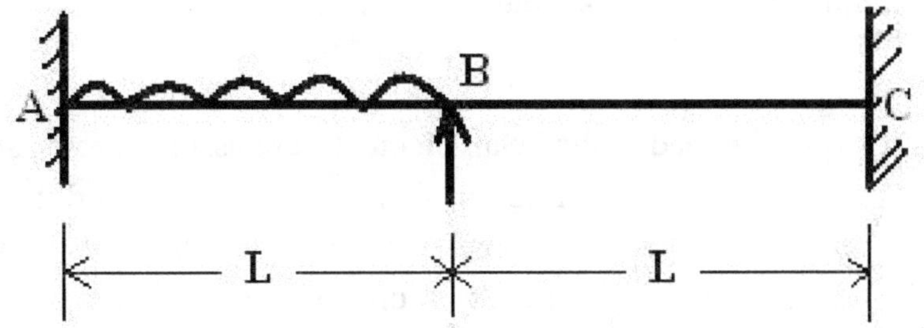

2.2 LINEAR STATIC ANALYSIS(BAR ELEMENT)

Most structural analysis problems can be treated as linear static problems, based on the following assumptions

1. Small deformations (loading pattern is not changed due to the deformed shape)
2. Elastic materials (no plasticity or failures)
3. Static loads (the load is applied to the structure in a slow or steady fashion)

Linear analysis can provide most of the information about the behavior of a structure, and can be a good approximation for many analyses. It is also the bases of nonlinear analysis in most of the cases.

2.3 BEAM ELEMENT

A beam element is defined as a long, slender member (one dimension is much larger than the other two) that is subjected to vertical loads and moments, which produce vertical displacements and rotations. The degrees of freedom for a beam element are a vertical displacement and a rotation at each node, as opposed to only an horizontal displacement at each node for a truss element.

Degrees of Freedom

Degrees of freedom are defined as the number of independent coordinates necessary to specify the configuration of a system. The degrees of freedom for a general situation consists of three translations in the *x, y,* and *z* directions and three rotations about the *x, y,* and *z* axes. A one-dimensional beam element has four degrees of freedom, which include, a vertical displacement (v) and a rotation (θ) at each node.

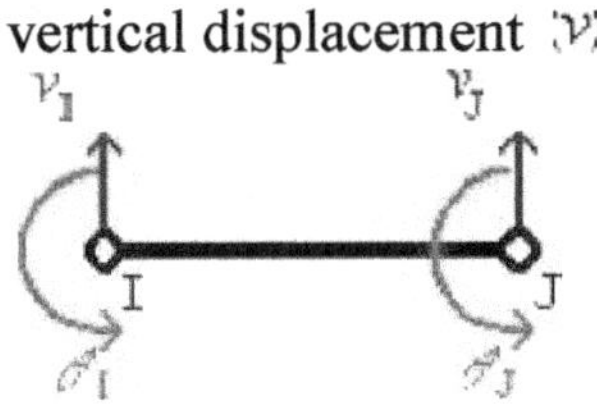

Assumptions

Nodal Forces and Moments

Forces and moments can only be applied at the nodes of the beam element, not between the nodes. The nodal forces and moments, f_E, are related to the nodal displacements and rotations, v_E through the element stiffness matrix, K_E.

Constant Load

The loads that are applied to the beam element are assumed to be static and not to vary over the time period being considered, this assumption is only valid if the rate of change of the force is much less than the applied force ($F >> dF/dt$). If the loads vary significantly, (if the variation in load is not much less than the applied force) then the problem **must be** considered as dynamic.

Weightless Member

The weight (W) of the beam is neglected, if it is much less than the total resultant forces (F) acting on the beam. If the weight of the beam is not neglected, then its effects must be represented as vertical forces acting at the nodes, by dividing up the weight and lumping it at the nodes, proportionally according to it's placement along the beam.

Prismatic Member

The beam element is assumed to have a constant cross-section, which means that the cross-sectional area and the moment of inertia will both be constant (i.e., the beam element is a prismatic member). If a beam is stepped, then itmust be divided up into sections of constant cross-section, in order to obtain an exact solution. If a beam is tapered, then the beam can be approximated by using many small beam elements, each having the same cross-section as the middle of the tapered length it is approximating. The more sections that are used to approximate a tapered beam, the more accurate the solution will be.

The moment of inertia is a geometric property of a beam element, which describes the beams resistance to bending and is assumed to be constant through the length of the element. The moment of inertia can be different along different axes if the beam element is not symmetric, we use the moment of inertia (I) of the axis about which the bending of the beam occurs

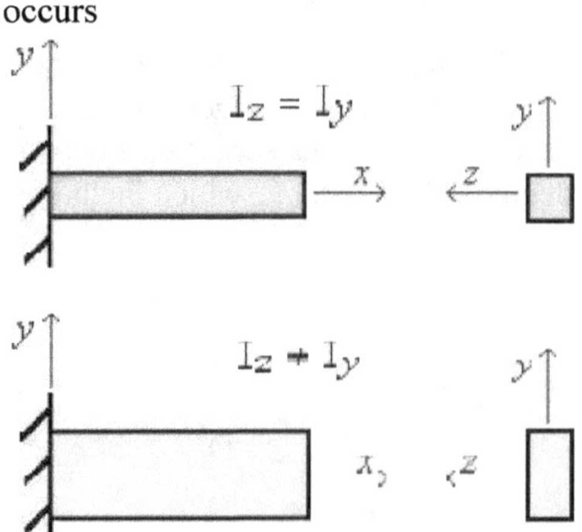

Where (I_z) refers to the moment of inertia, resisting bending about the "z" axis and (I_y) about the "y" axis.

The Beam Element is a Slender Member

A beam is assumed to be a slender member, when it's length (L) is more than 5 times as long as either of it's cross-sectional dimensions (d) resulting in (d/L<.2). A beam must be slender, in order for the beam equations to apply, that **were used to derive our FEM** equations.

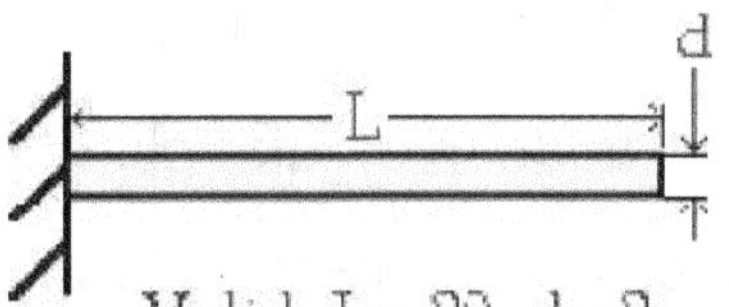

Valid L=20, d=2
d/L=.1

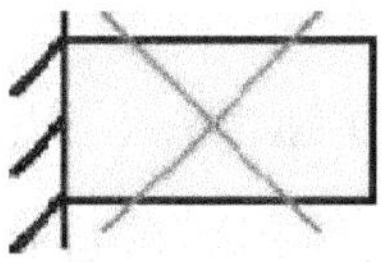

Invalid L=10, d=5
d/L=.5

The Beam Bends without Twisting.

It is assumed that the cross-section of the beam is symmetric about the plane of bending (*x-y* plane in this case) and will undergo **symmetric bending (where no twisting of** the beam occurs during the bending process). If the beam is not symmetric about this plane, then the beam will twist during bending and the situation will no longer be one-dimensional and must be approached as an unsymmetric bending problem (where the beam twists while bending) in order to obtain a correct solution.

Cross Section Remains Plane

When a beam element bends, it is assumed that it will deflect uniformly, thus the cross section will move uniformly and **remain plane to** the beam centerline. In other words, plane sections remain plane and normal to the x axis before and after bending.

Axially Rigid

The one-dimensional beam element is assumed to be axially rigid, meaning that there will be no axial displacement (u) along the beams centriodal axis. This implies that forces will only be applied perpendicular to the beams centriodal axis. The one-dimensional beam element can be used only when the degrees of freedom are limited to vertical displacements (perpendicular to the beams centriodal axis) and rotations in one plane. If axial displacements are present then a **one-dimensional** bar element **must be superimposed wiith the** one-dimensional beam element in order to obtain a valid solution.

Homogenous Material

A beam element has the same material composition throughout and therefore the same mechanical properties at every position in the material. Therefore, the modulus of elasticity E is constant throughout the beam element. A member in which the material properties varies from one point to the next in the member is called inhomogenous (non-homogenous). If a beam is composed of different types of materials, then it must be divide up into elements that are each of a single homogeneous material, otherwise the solution will not be exact.

[18] Cook RD, Malkus DS, and Plesha ME (1989) Concepts and Applications of Finite Element Analysis. 3rd edition, Wiley Publishing Co.

[19] Carpenter N, Stolarski H, and Belytschko T (1986) Improvements in 3-node triangular shell elements.International Journal for Numerical Methods in Engineering, 23(9): 1643-1667.

[20] Batoz JL, Bathe K-J, and Ho LW (1980) A study of three-node triangular plate bending elements.International Journal for Numerical Methods in Engineering, 15(12): 1771-1812.

[21] Lang S (1973) Real Analysis. Reading MA, Addison-Wesley.

[22] Murnaghan FD (1951) Finite Deformation of an Elastic Solid. Wiley, New York.

[23] Durran DR (1998) Numerical Methods for Wave Equations in Geophysical Fluid Dynamics. New York, NY, Springer-Verlag.

[24] Strang G, and Fix GJ (1973) An Analysis of the finite element method. Englewood Cliffs, NJ, Prentice Hall Series in Automatic Computation.

[25] Brooks AN, and Hughes TJR (1982) Streamline upwind/Petrov-Galerkin formulations for convection dominated flows with particular emphasis on the incompressible Navier-Stokes equations. Computer Methods in Applied Mechanics and Engineering, 32(1-3): 199-259.

[26] Carette JC, Deconinck H, Paillere H, Roe PL (1995) Multidimensional upwinding: its relation to finite elements. International Journal for Numerical Methods in Fluids, 20(8-9): 935-955.

[27] Warming RF, Beam RM, Hyett BJ (1975) Diagonalization and simultaneous symmetrization of the gas-dynamics matrices. Mathematics of Computation, 29(132): 1037-1045.

Isotropic Material

A beam element has the same mechanical and physical properties in all directions, i.e., they are independent of direction. For instance, cutting out three tensile test specimens, one in the *x*-direction, one in the *y*-**direction and the other oriented 45 degrees in the** *x-y* plane, a tension test on each specimen, will result in the same value for the modulus of elasticity (E), yield strength (σ_Y) and ultimate strength (σ_U). Most metals are considered isotropic. In contrast fibrous materials, such as **wood, typically have properties that are** directionaly dependant and are generally considered anisotropic (not isotropic).

The Proportional Limit is not Exceeded

It is assumed that the beam element is initially straight and unstressed. It is also assumed that the material does not yield, therefore the beam will be straight after the load is released. These assumpt**ions mean that the beam must be made of an elastic material, one** which will return to it's original size and shape when all loads are removed, if not stressed past the materials elastic or proportional limit. It is also assumed that the beam is not stressed past the proportional limit, at which point the beam will take a permanent set and will not fully return to it's original size and shape, when all loads are removed. Below the proportional limit an elastic material is in the linear elastic range, where the strain (ε) varies linearly with the applied load and the stress ( ) (σ) varies linearly according to: $\sigma = E\varepsilon$, where E is the modulus of elasticity.

Rigid Body Modes for the One-Dimensional Beam Element

Rigid body motion occurs when forces and/or moments are applied to an unrestrained mesh (body), resulting in motion that occurs without any deformations in the entire mesh (body). Since no strains (deformations) occur during rigid body motion, there can be no stresses developed in the mesh. In order to obtain a unique FEM solution, rigid body motion must be constrained. If rigid body motion is not constrained, the**n a singular system of** equations will result, since the determinate of the mesh stiffness matrix is equal to zero (i.e., $|\underline{K}| = 0$).

There are two rigid body modes for the one-dimensional beam element, a translation (displacement) only and a rotation only. These two rigid body modes can occur at the same time resulting in a displacement and a rotation simultaneously. In order to eliminate rigid body motion in a 1-D beam element (body), one must prescribe at least two nodal degrees of freedom (DOF), either two displacements or a displacement and a rotation. A DOF can be equal to zero or a non-zero known value, as long as the element **is restrained from rigid bod**y motion (deformation can take place when forces and moments are applied) .

For simplicity we will introduce the rigid body modes using a mesh composed of a singl**e element. If only** *translational rigid body motion* occurs, then the displacement at local node **I** will be equal to the displacement at local node **J.** Since the displacements are equal there is no strain developed in the element and the applied nodal forces cause the element to move in a rigid (non-deflected) vertical motion (which can be either up as shown below or it can be in the downward direction depending on the direction of the applied forces).

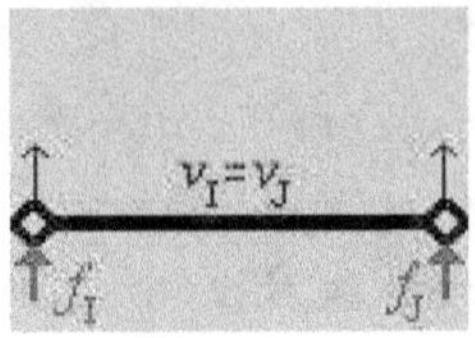

This rigid body mode can be suppressed by prescribing a vertical nodal displacement.

If rotational rigid body motion occurs, then the rotation at local node **I** will be equal to the rotation at local node **J** (i.e., in magnitude and direction). In this si**tuation the** nodal forces and/or moments applied to the element, cause the element to rotate as a rigid body (either clockwise as shown below or counterclockwise depending on the direction of the applied forces and/or moments).

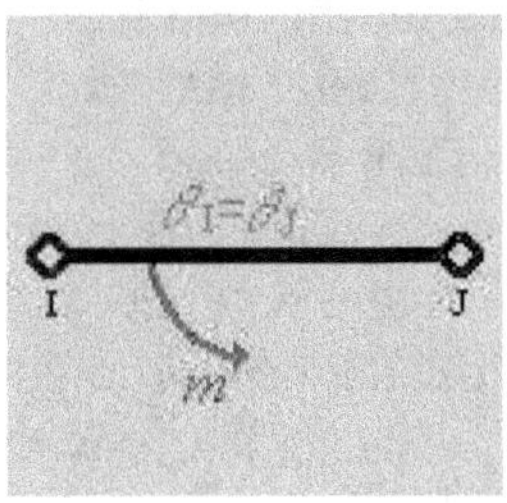

This rigid body mode can be suppressed by prescribing a nodal translation or rotation.

If translational *and* rotational rigid body motion occurs simultaneously then:

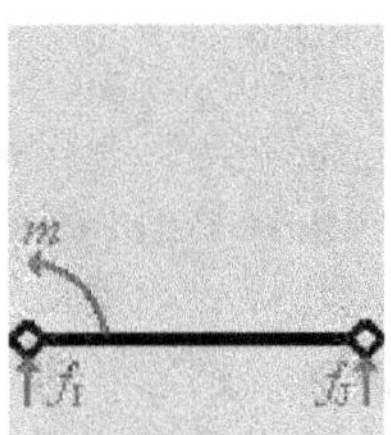

Simple Examples of Beam Problems with and without Rigid Body Motion

Case	**Stable/Unstable Structure**	**Rigid Body Mode(s) Present**	**Determinant of Mesh Stiffness Matrix**	**Equations**
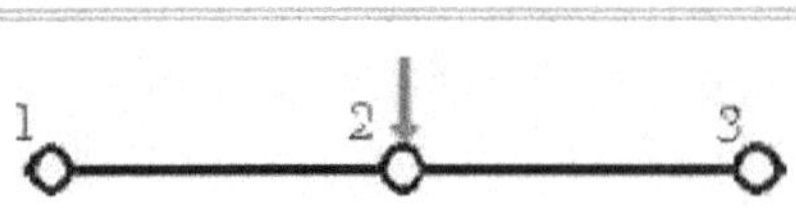	**Unstable**	θ **and** v	$\lvert K \rvert = 0$	**Dependent Equations**
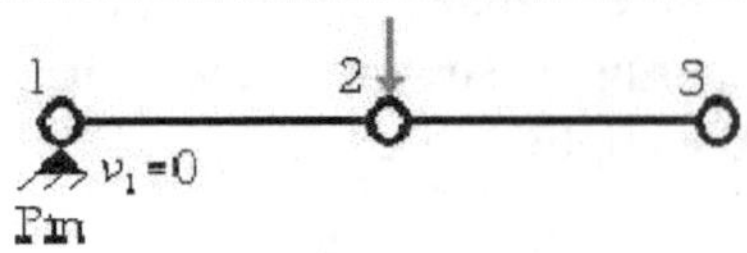	**Unstable**	θ	$\lvert K \rvert = 0$	**Dependent Equations**

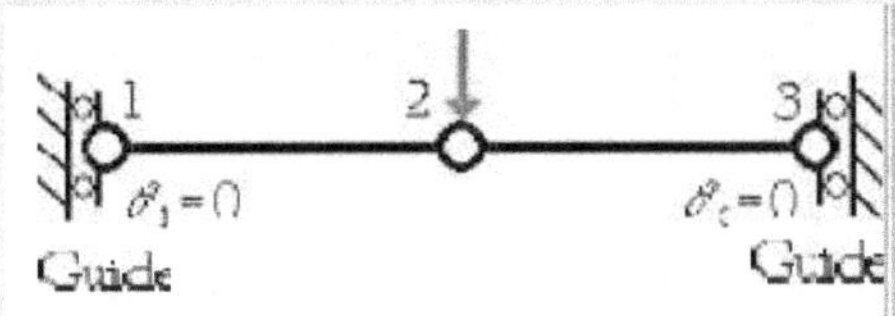	**Unstable**	v	$\lvert K \rvert = 0$	**Dependent Equations**
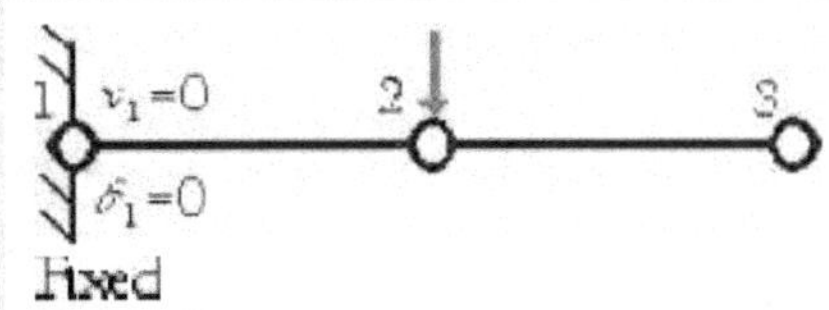	**Stable**	**None**	$\lvert K \rvert \neq 0$	**Independen t Equations**
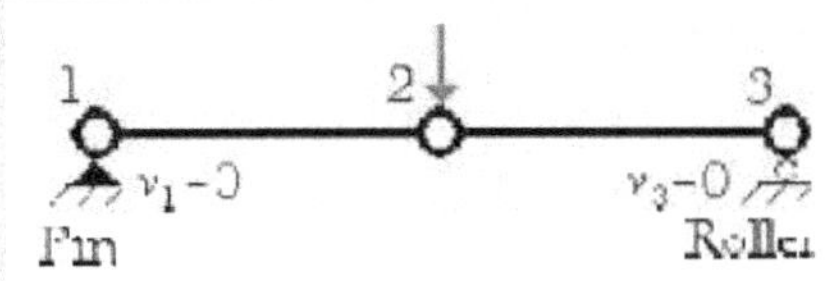	**Stable**	**None**	$\lvert K \rvert \neq 0$	**Independen t Equations**

2.4 1-D 2-NODED CUBIC BEAM ELEMENT MATRICES

A single 1-d 2-noded cubic beam element has two nodes, with two degrees of freedom at each node (one vertical displacement and one rotation or slope). There is a total of 4 dof and the displacement polynomial function assumed should have 4 terms, so we choose a cubic polynomial for the vertical deflection. Slope is a derivative of the vertical deflections.

The vertical displacement $$v = a + bx + cx^2 + dx^3 \qquad \text{.......................(1)}$$

The slope $$\theta = \frac{dv}{dx} = b + 2cx + 3dx^2 \qquad \text{.......................(2)}$$

Apply the boundary conditions

$$\text{at } x = 0, \quad v = v_1 \quad \Rightarrow \quad v_1 = a \quad \Rightarrow \quad a = v_1$$

$$\text{at } x = 0, \quad \theta = \theta_1 \quad \Rightarrow \quad \theta_1 = b \quad \Rightarrow \quad b = \theta_1$$

$$\left.\begin{aligned} \text{at } x = l, \quad v = v_2 \quad &\Rightarrow \quad v_2 = a + bl + cl^2 + dl^3 \\ \text{at } x = l, \quad \theta = \theta_2 \quad &\Rightarrow \quad \theta_2 = b + 2cl + 3dl^2 \end{aligned}\right\} \text{ solving } \quad \begin{aligned} c &= \frac{3}{l^2}(v_2 - v_1) - \frac{1}{l}(2\theta_1 + \theta_2) \\ d &= \frac{2}{l^3}(v_1 - v_2) + \frac{1}{l^2}(\theta_1 + \theta_2) \end{aligned}$$

Substituting the values of a, b, c and d in equation (1), and collecting the coefficients of v_1, θ_1, v_2, θ_2 we obtain

$$v = N_1 v_1 + N_2 \theta_1 + N_3 v_3 + N_4 \theta_2$$

where

$$N_1 = 1 - 3\frac{x^2}{l^2} + 2\frac{x^3}{l^3}, \quad N_2 = x - 2\frac{x^2}{l} + \frac{x^3}{l^2},$$

$$N_3 = 3\frac{x^2}{l^2} - 2\frac{x^3}{l^3}, \quad N_4 = -\frac{x^2}{l} + \frac{x^3}{l^2}$$

2.5 DEVELOPMENT OF ELEMENT EQUATION

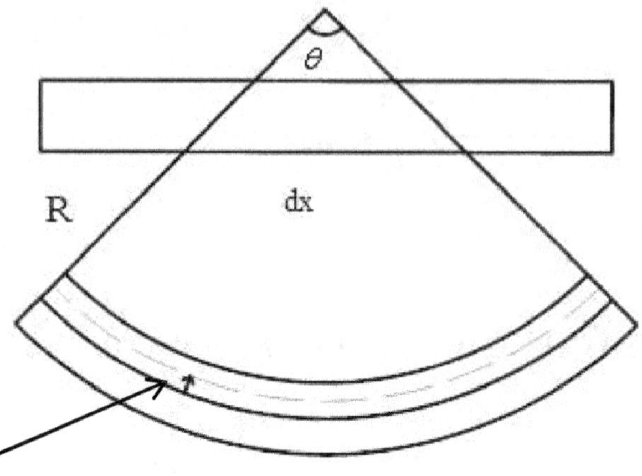

$$dx = R\theta$$

$$\epsilon_x = \frac{(R-y)\theta - R\theta}{R\theta} = \frac{-y}{R} = -y\frac{d^2y}{dx^2}$$

$$v = N_1 v_1 + N_2\theta_1 + N_3 v_2 + N_4\theta_2$$

$$\epsilon_x = -y\frac{d^2v}{dx^2} = -y\frac{d^2}{dx^2}\left(N_1 v_1 + N_2\theta_1 + N_3 v_2 + N_4\theta_2\right)$$

$$\epsilon_x = -y\underbrace{\begin{pmatrix}\frac{d^2N_1}{dx^2} & \frac{d^2N_2}{dx^2} & \frac{d^2N_3}{dx^2} & \frac{d^2N_4}{dx^2}\end{pmatrix}}_{[B]}\underbrace{\begin{Bmatrix}v_1\\ \theta_1\\ v_2\\ \theta_2\end{Bmatrix}}_{\{a\}}$$

$$\epsilon_x = [B]\{a\}$$

We Know that,

$$[K]=\int_v [B]^T D [B]\, dv$$

$$[K]=\int_{volume} (-y)\begin{pmatrix}\frac{d^2N_1}{dx^2}\\ \frac{d^2N_2}{dx^2}\\ \frac{d^2N_3}{dx^2}\\ \frac{d^2N_4}{dx^2}\end{pmatrix} E(-y)\begin{pmatrix}\frac{d^2N_1}{dx^2} & \frac{d^2N_2}{dx^2} & \frac{d^2N_3}{dx^2} & \frac{d^2N_4}{dx^2}\end{pmatrix} dv$$

$$[K]=E\int_v \int_0^l y^2 \begin{pmatrix} \left(\frac{d^2N_1}{dx^2}\right)^2 & \left(\frac{d^2N_1}{dx^2}\frac{d^2N_2}{dx^2}\right) & \left(\frac{d^2N_1}{dx^2}\frac{d^2N_3}{dx^2}\right) & \left(\frac{d^2N_1}{dx^2}\frac{d^2N_4}{dx^2}\right)\\ \left(\frac{d^2N_2}{dx^2}\frac{d^2N_1}{dx^2}\right) & \left(\frac{d^2N_2}{dx^2}\right)^2 & \left(\frac{d^2N_2}{dx^2}\frac{d^2N_3}{dx^2}\right) & \left(\frac{d^2N_2}{dx^2}\frac{d^2N_4}{dx^2}\right)\\ \left(\frac{d^2N_3}{dx^2}\frac{d^2N_1}{dx^2}\right) & \left(\frac{d^2N_3}{dx^2}\frac{d^2N_2}{dx^2}\right) & \left(\frac{d^2N_3}{dx^2}\right)^2 & \left(\frac{d^2N_3}{dx^2}\frac{d^2N_4}{dx^2}\right)\\ \left(\frac{d^2N_4}{dx^2}\frac{d^2N_1}{dx^2}\right) & \left(\frac{d^2N_4}{dx^2}\frac{d^2N_2}{dx^2}\right) & \left(\frac{d^2N_4}{dx^2}\frac{d^2N_3}{dx^2}\right) & \left(\frac{d^2N_4}{dx^2}\right)^2 \end{pmatrix} dA\,dx$$

Where, $\int_0^l y^2 dA = I$

$$[K]=EI\int_0^v \begin{pmatrix}\frac{d^2N_1}{dx^2}\\ \frac{d^2N_2}{dx^2}\\ \frac{d^2N_3}{dx^2}\\ \frac{d^2N_4}{dx^2}\end{pmatrix}\begin{pmatrix}\frac{d^2N_1}{dx^2} & \frac{d^2N_2}{dx^2} & \frac{d^2N_3}{dx^2} & \frac{d^2N_4}{dx^2}\end{pmatrix} dx$$

Where,

$$N_1 = 1-\frac{3x^2}{l^2}+\frac{2x^3}{l^3} \Rightarrow \frac{dN_1}{dx}=\frac{-6x}{l^2}+\frac{6x^2}{l^3} \Rightarrow \frac{d^2N_1}{dx^2}=\frac{-6}{l^2}+\frac{12x}{l^3}$$

$$N_2 = x-\frac{2x^2}{l}+\frac{x^3}{l^2} \Rightarrow \frac{dN_2}{dx}=1-\frac{4x}{l}+\frac{3x^2}{l^2} \Rightarrow \frac{d^2N_2}{dx^2}=\frac{-4}{l}+\frac{6x}{l^2}$$

$$N_3 = \frac{3x^2}{l^2} - \frac{3x^3}{l^3} \Rightarrow \frac{dN_3}{dx} = \frac{6x}{l^2} - \frac{6x^2}{l^3} \Rightarrow \frac{d^2N_3}{dx^2} = \frac{6}{l^2} - \frac{12x}{l^3}$$

$$N_4 = \frac{x^3}{l^2} - \frac{x^2}{l} \Rightarrow \frac{dN_4}{dx} = \frac{3x}{l^2} - \frac{2x}{l} \Rightarrow \frac{d^2N_4}{dx^2} = \frac{6x}{l^2} - \frac{2}{l}$$

$$[K] = EI\int_0^l \begin{pmatrix} \left(\frac{-6}{l^2} + \frac{12x}{l^3}\right) \\ \left(\frac{-4}{l} + \frac{6x}{l^2}\right) \\ \left(\frac{6}{l^2} - \frac{12x}{l^3}\right) \\ \left(\frac{6x}{l^2} - \frac{2}{l}\right) \end{pmatrix} \left(\left(\frac{-6}{l^2} + \frac{12x}{l^3}\right) \left(\frac{-4}{l} \frac{6x}{l^2}\right) \left(\frac{6}{l^2} - \frac{12x}{l^3}\right) \left(\frac{6x}{l^2} \frac{2}{l}\right)\right) dx$$

$$K_{11} = EI\int_0^l \left(\frac{-6}{l^2} + \frac{12x}{l^3}\right)\left(\frac{-6}{l^2} + \frac{12x}{l^3}\right) dx$$

$$K_{11} = EI\int_0^l \left(\frac{-6l + 12x}{l^3}\right)\left(\frac{-6l + 12x}{l^3}\right) dx$$

$$K_{11} = EI\int_0^l \left(\frac{36l^2 - 72xl - 72xl + 144x^2}{l^6}\right) dx$$

$$K_{11} = EI\int_0^l \left(\frac{36l^2}{l^6} - \frac{144xl}{l^6} + \frac{144x^2}{l^6}\right) dx$$

$$K_{11} = EI\left[\frac{36xl^2}{l^6} - \frac{144x^2l}{2l^6} + \frac{144x^3}{3l^6}\right]_o^l$$

$$K_{11} = EI\left[\frac{-36}{l^3} - \frac{72}{l^3} + \frac{48}{l^3}\right]$$

$$\boxed{K_{11} = \frac{12EI}{l^3}}$$

$$K_{12} = EI\int_0^l \left(\frac{-6}{l^2} + \frac{12x}{l^3}\right)\left(\frac{-4}{l} + \frac{6x}{l^2}\right) dx$$

$$K_{12} = EI\int_0^l \left(\frac{-6l + 12x}{l^3}\right)\left(\frac{-4l + 6x}{l^2}\right)dx$$

$$K_{12} = EI\int_0^l \left(\frac{24l^2 - 48xl - 36xl + 72x^2}{l^5}\right)dx$$

$$K_{12} = EI\int_0^l \left(\frac{24l^2}{l^5} - \frac{84xl}{l^5} + \frac{72x^2}{l^5}\right)dx$$

$$K_{12} = EI\int_0^l \left[\frac{24xl^2}{l^5} - \frac{84x^2 l}{2l^5} + \frac{72x^3}{3l^5}\right]_0^l$$

$$K_{12} = EI\left[\frac{24}{l^2} - \frac{42}{l^2} + \frac{24}{l^2}\right]$$

$$\boxed{K_{12} = \frac{6EI}{l^2}}$$

$$K_{13} = EI\int_0^l \left(\frac{-6}{l^2} + \frac{12x}{l^3}\right)\left(\frac{6}{l^2} + \frac{-12x}{l^3}\right)dx$$

$$K_{13} = EI\int_0^l \left(\frac{-6l + 12x}{l^3}\right)\left(\frac{6l - 12x}{l^3}\right)dx$$

$$K_{13} = EI\int_0^l \left(\frac{-36l^2 + 72xl + 72xl - 144x^2}{l^6}\right)dx$$

$$K_{13} = EI\int_0^l \left[\frac{-36xl^2}{l^6} + \frac{144x^2 l}{2l^6} - \frac{144x^3}{3l^6}\right]_0^l$$

$$K_{13} = EI\left[\frac{-36 + 72 - 48}{l^3}\right]$$

$$\boxed{K_{13} = \frac{-12EI}{l^3}}$$

$$K_{14} = EI\int_0^l \left(\frac{-6}{l^2} + \frac{12x}{l^3}\right)\left(\frac{6x}{l^2} - \frac{2}{l}\right)dx$$

$$K_{14} = EI\int_0^l \left(\frac{-6l + 12x}{l^3}\right)\left(\frac{6x - 2l}{l^2}\right)dx$$

$$K_{14} = EI\int_0^l \left(\frac{12l^2 - 36xl - 24xl + 72x^2}{l^5} \right) dx$$

$$K_{14} = EI\int_0^l \left(\frac{12l^2}{l^5} - \frac{60xl}{l^5} + \frac{72x^2}{l^5} \right) dx$$

$$K_{14} = EI\int_0^l \left[\frac{12xl^2}{l^5} - \frac{60x^2 l}{2l^5} + \frac{72x^3}{3l^5} \right]_0^l$$

$$K_{14} = EI\left[\frac{-30 + 12 + 24}{l^2} \right]$$

$$\boxed{K_{14} = \frac{6EI}{l^2}}$$

$$K_{21} = EI\int_0^l \left(\frac{-4}{l} + \frac{6x}{l^2} \right)\left(\frac{-6}{l^2} + \frac{12x}{l^3} \right) dx$$

$$\boxed{K_{21} = \frac{6EI}{l^2}}$$

$$K_{22} = EI\int_0^l \left(\frac{-4}{l} + \frac{6x}{l^2} \right)\left(\frac{-4}{l} + \frac{6x}{l^2} \right) dx$$

$$K_{22} = EI\int_0^l \left(\frac{-4l + 6x}{l^2} \right)\left(\frac{-4l + 6x}{l^2} \right) dx$$

$$K_{22} = EI\int_0^l \left(\frac{16l^2 - 24xl - 24xl + 36x^2}{l^4} \right) dx$$

$$K_{22} = EI\int_0^l \left(\frac{16l^2}{l^4} - \frac{48xl}{l^4} + \frac{36x^2}{l^4} \right) dx$$

$$K_{22} = EI\int_0^l \left[\frac{16xl^2}{l^4} - \frac{48x^2 l}{2l^4} + \frac{36x^3}{3l^4} \right]_0^l$$

$$K_{22} = EI\left[\frac{16 - 24 + 12}{l} \right]$$

$$\boxed{K_{22} = \frac{4EI}{l}}$$

$$K_{23} = EI\int_0^l \left(\frac{-4}{l} + \frac{6x}{l^2}\right)\left(\frac{6}{l^2} - \frac{12x}{l^3}\right)dx$$

$$K_{23} = EI\int_0^l \left(\frac{-4l + 6x}{l^2}\right)\left(\frac{6l - 12x}{l^3}\right)dx$$

$$K_{23} = EI\int_0^l \left(\frac{-24l^2 + 36xl + 48xl - 72x^2}{l^5}\right)dx$$

$$K_{23} = EI\int_0^l \left(\frac{-24l^2 + 84xl - 72x^2}{l^5}\right)dx$$

$$K_{23} = EI\int_0^l \left[\frac{-24xl^2}{l^5} + \frac{84x^2l}{2l^5} - \frac{72x^3}{3l^5}\right]_0^l$$

$$K_{23} = EI\left[\frac{-24 + 42 - 24}{l^2}\right]$$

$$\boxed{K_{23} = \frac{-6EI}{l^2}}$$

$$K_{24} = EI\int_0^l \left(\frac{-4}{l} + \frac{6x}{l^2}\right)\left(\frac{6x}{l^2} - \frac{2}{l}\right)dx$$

$$K_{24} = EI\int_0^l \left(\frac{-4l + 6x}{l^2}\right)\left(\frac{6x - 2l}{l^2}\right)dx$$

$$K_{24} = EI\int_0^l \left(\frac{8l^2 - 24xl - 12xl + 36x^2}{l^4}\right)dx$$

$$K_{24} = EI\int_0^l \left(\frac{8l^2 - 36xl + 36x^2}{l^4}\right)dx$$

$$K_{24} = EI\int_0^l \left[\frac{8xl^2}{l^4} - \frac{36x^2l}{2l^4} + \frac{36x^3}{3l^4}\right]_0^l$$

$$K_{24} = EI\left[\frac{-18 + 12 + 8}{l}\right]$$

$$\boxed{K_{24} = \frac{2EI}{l}}$$

$$K_{31} = EI\int_0^l \left(\frac{6}{l^2} + \frac{-12x}{l^3}\right)\left(\frac{-6}{l^2} + \frac{12x}{l^3}\right)dx$$

$$\boxed{K_{31} = \frac{-12EI}{l^3}}$$

$$K_{32} = EI\int_0^l \left(\frac{6}{l^2} - \frac{12x}{l^3}\right)\left(\frac{-4}{l} + \frac{6x}{l^2}\right)dx$$

$$\boxed{K_{32} = \frac{-6EI}{l^2}}$$

$$K_{33} = EI\int_0^l \left(\frac{6}{l^2} - \frac{12x}{l^3}\right)\left(\frac{6}{l^2} - \frac{12x}{l^3}\right)dx$$

$$K_{33} = EI\int_0^l \left(\frac{6l - 12x}{l^3}\right)\left(\frac{6l - 12x}{l^3}\right)dx$$

$$K_{33} = EI\int_0^l \left(\frac{36l^2 - 72xl - 72xl + 144x^2}{l^6}\right)dx$$

$$K_{33} = EI\int_0^l \left(\frac{36l^2 - 144xl + 144x^2}{l^6}\right)dx$$

$$K_{33} = EI\int_0^l \left[\frac{36\,xl^2}{l^6} - \frac{144x^2l}{2l^6} + \frac{144\,x^3}{3l^6}\right]_0^l$$

$$K_{33} = EI\left[\frac{36 - 72 + 48}{l^3}\right]$$

$$\boxed{K_{33} = \frac{12EI}{l^3}}$$

$$K_{34} = EI\int_0^l \left(\frac{6}{l^2} - \frac{12x}{l^3}\right)\left(\frac{6x}{l^2} - \frac{2}{l}\right)dx$$

$$K_{34} = EI\int_0^l \left(\frac{6l - 12x}{l^3}\right)\left(\frac{6x - 2l}{l^2}\right)dx$$

$$K_{34} = EI\int_0^l \left(\frac{-12l^2 + 24xl + 36xl - 72x^2}{l^5}\right)dx$$

$$K_{34} = EI\int_0^l \left(\frac{-12l^2 + 60xl - 72x^2}{l^5} \right) dx$$

$$K_{34} = EI\int_0^l \left[\frac{-12xl^2}{l^5} + \frac{60x^2 l}{2l^5} - \frac{72x^3}{3l^5} \right]_0^l$$

$$K_{34} = EI\left[\frac{-12+30-24}{l^2} \right]$$

$$\boxed{K_{34} = \frac{-6EI}{l^2}}$$

$$K_{41} = EI\int_0^l \left(\frac{6x}{l^2} - \frac{2}{l} \right)\left(\frac{-6}{l^2} + \frac{12x}{l^3} \right) dx$$

$$\boxed{K_{41} = \frac{6EI}{l^2}}$$

$$K_{42} = EI\int_0^l \left(\frac{6x}{l^2} - \frac{2}{l} \right)\left(\frac{-4}{l} + \frac{6x}{l^2} \right) dx$$

$$\boxed{K_{42} - \frac{2EI}{l}}$$

$$K_{43} = EI\int_0^l \left(\frac{6x}{l^2} - \frac{2}{l} \right)\left(\frac{6}{l^2} - \frac{12x}{l^3} \right) dx$$

$$\boxed{K_{43} = \frac{-6EI}{l^2}}$$

$$K_{44} = EI\int_0^l \left(\frac{6x}{l^2} - \frac{2}{l} \right)\left(\frac{6x}{l^2} - \frac{2}{l} \right) dx$$

$$K_{44} = EI\int_0^l \left(\frac{6x-2l}{l^2} \right)\left(\frac{6x-2l}{l^2} \right) dx$$

$$K_{44} = EI\int_0^l \left(\frac{4l^2 - 12xl - 12xl + 36x^2}{l^4} \right) dx$$

$$K_{44} = EI\int_0^l \left(\frac{4l^2 - 24xl + 36x^2}{l^4} \right) dx$$

$$K_{44} = EI\int_0^l \left[\frac{4xl^2}{l^4} - \frac{24x^2 l}{2l^4} + \frac{36x^3}{3l^4}\right]_0^l$$

$$K_{44} = EI\left[\frac{12+4-12}{l}\right]$$

$$\boxed{K_{44} = \frac{4EI}{l}}$$

Therefore $[K]$ is

$$[K] = EI\begin{pmatrix} \frac{12}{l^3} & \frac{6}{l^2} & \frac{-12}{l^3} & \frac{6}{l^2} \\ \frac{6}{l^2} & \frac{4}{l} & \frac{-6}{l^2} & \frac{2}{l} \\ \frac{-12}{l^3} & \frac{-6}{l^2} & \frac{12}{l^3} & \frac{-6}{l^2} \\ \frac{6}{l^2} & \frac{2}{l} & \frac{-6}{l^2} & \frac{4}{l} \end{pmatrix}$$

2.6 BEAM ELEMENT

A beam is a long, slender structural member generally subjected to transverse loading that produces significant bending effects as opposed to twisting or axial effects. An elemental length of a long beam subjected to arbitrary loading is considered for analysis. For this elemental beam length L, we assign two points of interest, i.e., the ends of the beam, which become the nodes of the beam element. The bending deformation is measured as a transverse (vertical) displacement and a rotation (slope). Hence, for each node, we have a vertical displacement and a rotation (slope) – two degrees of freedom at each node. For a single 2-noded beam element, we have a total of 4 degrees of freedom. The associated "forces" are shear force and bending moment at each node.

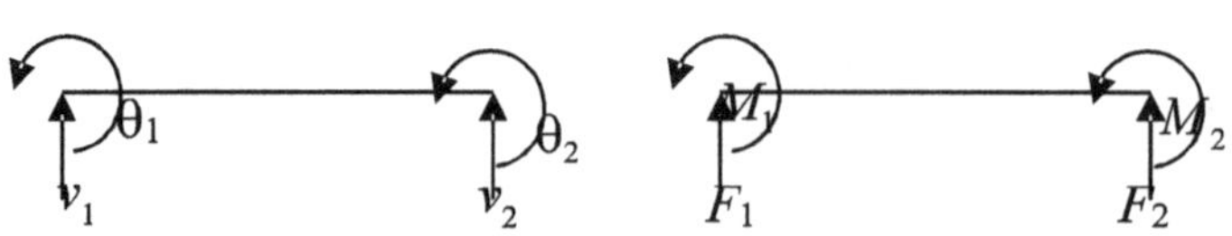

Nodal "displacements" Nodal "forces"

1st degree of freedom	vertical displacement at node i	1	v_i or v_1	corres-pond-ing to	shear force at node i	F_i or F_1	1
2nd degree of freedom	slope or rotation at node i	2	θ_i or θ_1		bending moment at node i	M_i or M_1	2
3rd	vertical	3	v_j or v_2		shear force at node	F_j or F_2	3

degree of freedom	displacement at node j				i		
4th degree of freedom	slope or rotation at node j	4	θ_j or θ_2		bending moment at node j	M_j or M_2	4

The stiffness term k_{ij} indicates the force (or moment) required at i to produce a unit deflection (or rotation) at j, while all other degrees of freedom are kept zero.

Sign conventions followed

Upward forces are positive and upward displacements are positive.

Counter-clockwise moments are positive and counter-clockwise rotations are positive.

Formulae required – cantilever beam subjected to concentrated load and moment.

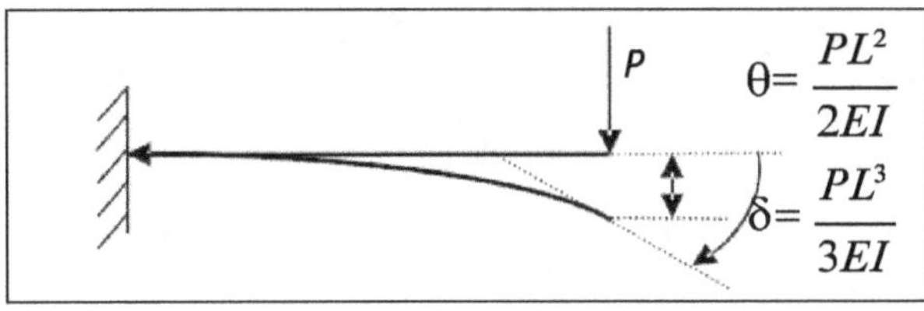

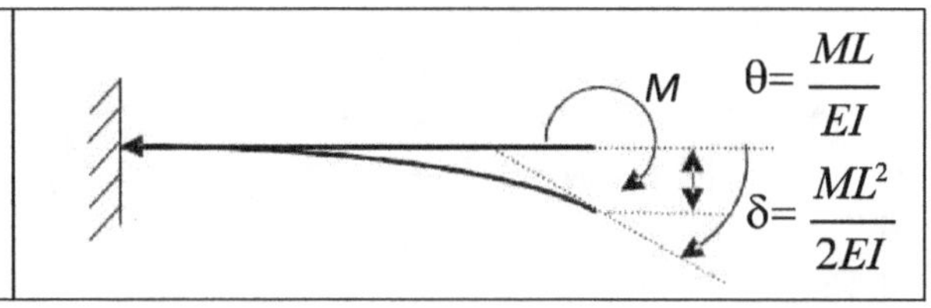

2.6.1 ELEMENT MATRICES AND VECTORS

Derivation of first column of stiffness matrix $v_1 = 1, \ \theta_1 = v_2 = \theta_2 = 0$, i.e., allow the first degree of freedom to occur and arrest all other DoF. (The deformed configuration is shown in Figure 2).

Initially you have a horizontal beam element. Since $v_2 = \theta_2 = 0$, we can fix node j. To produce an upward deflection at node i (i.e., allowing first degree of freedom to occur), apply an upward force k_{11} (first suffix indicates the force or moment DoF and the second suffix indicates the displacement or rotational DoF). $v_1 = \frac{k_{11} L^3}{3EI}$ upwards. Refer table for displacement DoF number and force DoF number. Now the beam configuration is given by Figure 1. We can observe from the figure that the slope at node i is not zero. To make the slope at i equal to zero, we need to apply a counter-clockwise moment k_{21}. Refer Figure 2. But this moment k_{21} will produce a downward deflection $\frac{k_{21} L^2}{2EI}$ at node i. Refer Figure

3. In order to have a resultant unit upward displacement at node i, upward displacement produced by force k_{11} must be greater than the downward displacement produced by the moment k_{21}. i.e., $\frac{k_{11} L^3}{3EI} - \frac{k_{21} L^2}{2EI} = 1$ .....(1). At the same time, the negative slope produced at node i by the force k_{11} must be cancelled by the positive slope produced by the moment k_{21}. i.e., $\frac{k_{11} L^2}{2EI} = \frac{k_{21} L}{EI}$ ....(2). Solving these two equations, k_{11} and k_{21} are found. The fixed end

reaction force and the reaction moment are assumed to be acting upwards and counterclockwise, respectively. Now use force equilibrium equation to find fixed end reaction force k_{31} .... $\left(\sum F_y = 0 \Rightarrow k_{11} + k_{31} = 0\right)$ and moment equilibrium equation about node i to find fixed end reaction moment k_{41} .... $\left(\sum M_i = 0 \Rightarrow k_{21} + k_{31}L + k_{41} = 0\right)$.

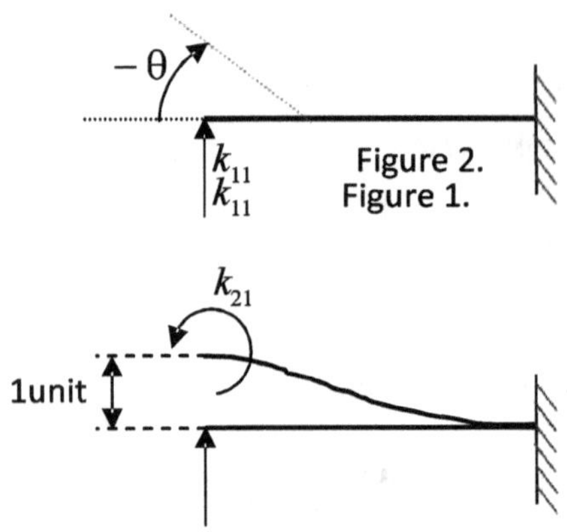

Figure 2.
Figure 1.

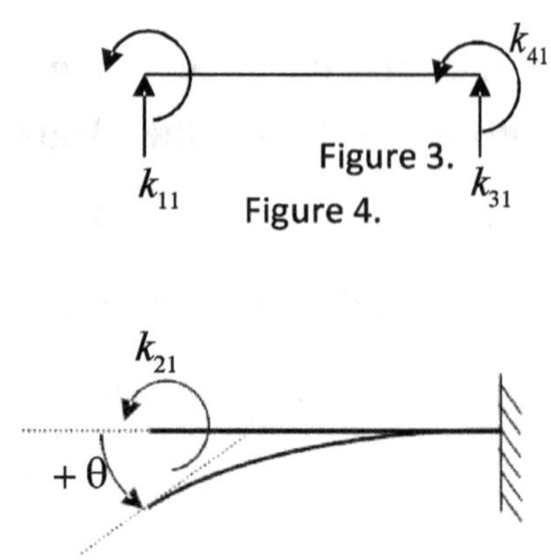

Figure 3.
Figure 4.

$$\begin{bmatrix} k_{11} \\ k_{21} \\ k_{31} \\ k_{41} \end{bmatrix} = \begin{bmatrix} \dfrac{12EI}{L^3} \\ \dfrac{6EI}{L^2} \\ -\dfrac{12EI}{L^3} \\ \dfrac{6EI}{L^2} \end{bmatrix}$$

Derivation of second column of stiffness matrix: $v_1 = 0,\ \theta_1 = 1, v_2 = \theta_2 = 0$, i.e., allow the second degree of freedom to occur and arrest all other DoF. (The deformed configuration is shown in Figure 2).

Initially you have a horizontal beam element. Since $v_2 = \theta_2 = 0$, we can fix node j. To produce a counterclockwise (positive) rotation or slope at node i (i.e., allowing second degree of freedom to occur), apply a counterclockwise moment k_{22}. $\theta_1 = \frac{k_{22}L}{EI}$. Refer Figure 1. This moment k_{22} will produce a downward deflection $\frac{k_{22}L^2}{2EI}$. This downward deflection should be canceled by applying an upward force k_{12} at node i. The upward deflection produced by k_{12} is $\frac{k_{12}L^3}{3EI}$. Refer Figure 2. Equating these two deflections $\frac{k_{22}L^2}{2EI} = \frac{k_{12}L^3}{3EI}$ ...(1) But this upward force k_{12} will also produce a negative slope at node i which is $\frac{k_{12}L^2}{2EI}$. Refer Figure 3. Hence the rotation produced by k_{22} should be greater than that produced by k_{12} so that the resultant rotation is 1 radians. $\frac{k_{22}L}{EI} - \frac{k_{12}L^2}{2EI} = 1$ ....(2). Solving these two equations, k_{12} and k_{22} are found. The fixed end reaction force and the reaction moment are assumed to be acting upwards and counterclockwise, respectively. Now use force equilibrium equation to find fixed end reaction force k_{32} ... $\left(\sum F_y = 0 \Rightarrow k_{12} + k_{32} = 0\right)$ and moment equilibrium equation about node i to find fixed end reaction moment k_{42} .... $\left(\sum M_i = 0 \Rightarrow k_{22} + k_{32}L + k_{42} = 0\right)$.

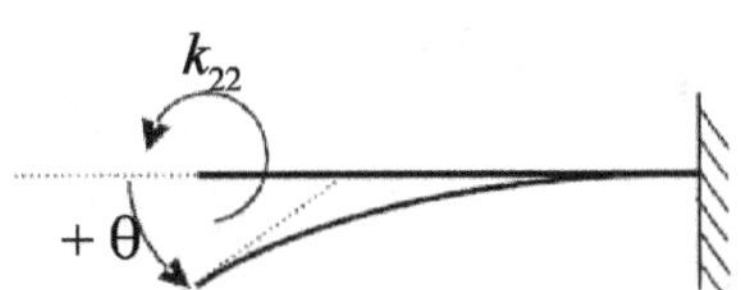

Figure 1.

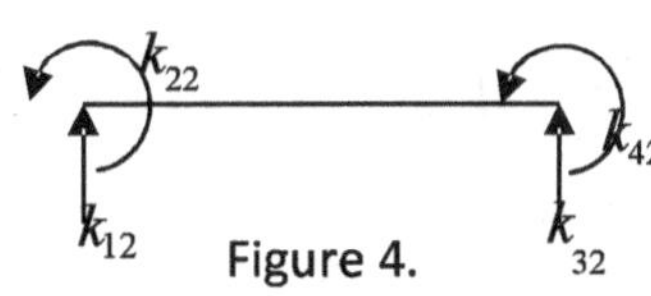

Figure 4.

$$\begin{bmatrix} k_{12} \\ k_{22} \\ k_{32} \\ k_{42} \end{bmatrix} = \begin{bmatrix} \frac{6EI}{L^2} \\ \frac{4EI}{L} \\ -\frac{6EI}{L^2} \\ \frac{2EI}{L} \end{bmatrix}$$

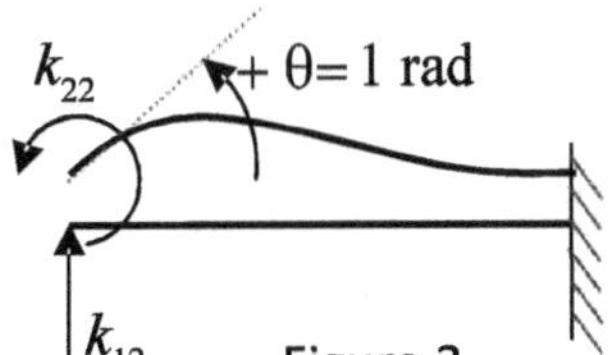

Figure 2.

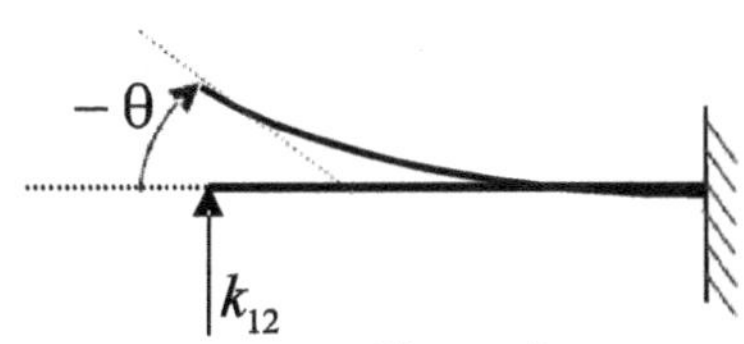

Figure 3.

Derivation of third column of stiffness matrix: $v_1 = 0,\ \theta_1 = 0, v_2 = 1, \theta_2 = 0$, i.e., allow the third degree of freedom to occur and arrest all other DoF. (The deformed configuration is shown in Figure 2).

Initially you have a horizontal beam element. Since $v_1 = \theta_1 = 0$, we can fix node i. To produce an upward deflection at node j (i.e., allowing third degree of freedom to occur), apply an upward force k_{33}.

$v_2 = \dfrac{k_{33} L^3}{3EI}$ upwards. Now the beam configuration is given by Figure 1. We can observe from the figure that the slope at node j is not zero. To make the slope at j equal to zero, we need to apply a clockwise moment k_{43}. Refer Figure 2. But this moment k_{43} will produce a downward deflection $\dfrac{k_{43} L^2}{2EI}$ at node j. Refer Figure 3. In order to have a resultant unit upward displacement at node j, upward displacement produced by force k_{33} must be greater than the downward displacement produced by the moment k_{43}. i.e., $\dfrac{k_{33} L^3}{3EI} - \dfrac{k_{43} L^2}{2EI} = 1$ .....(1). At the same time, the positive slope produced at node j by the force k_{33} must be cancelled by the negative slope produced by the moment k_{43}. i.e., $\dfrac{k_{33} L^2}{2EI} = \dfrac{k_{43} L}{EI}$ ....(2). Solving these two equations, k_{33} and k_{43} are found. The fixed end reaction force and the reaction moment are assumed to be acting upwards and counterclockwise, respectively. Now use force equilibrium equation to find fixed end reaction force k_{13} ... $\left(\sum F_y = 0 \;\Rightarrow\; k_{13} + k_{33} = 0\right)$ and moment equilibrium equation about node i to find fixed end reaction moment k_{23} .... $\left(\sum M_i = 0 \;\Rightarrow\; k_{23} + k_{33}L + k_{43} = 0\right)$.

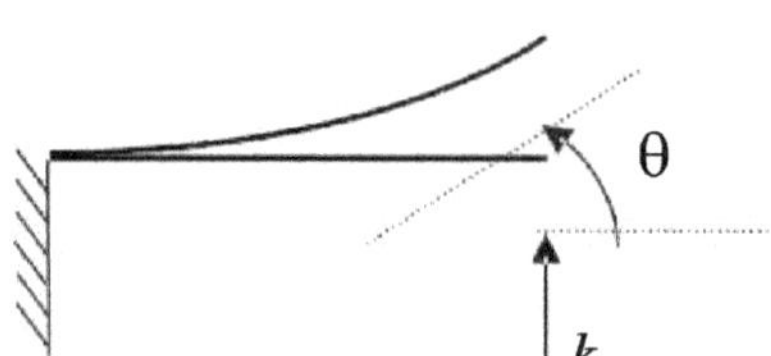

Figure 1.

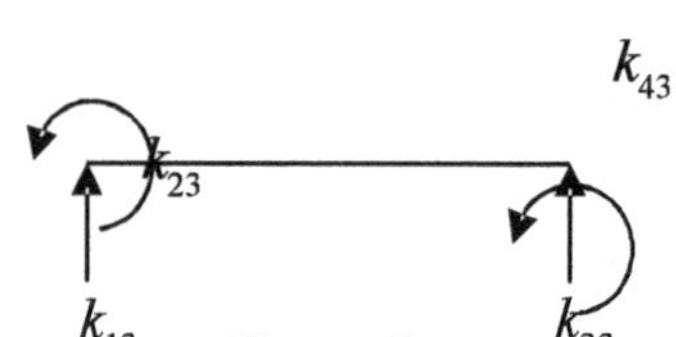

Figure 4.

$$\begin{bmatrix} k_{13} \\ k_{23} \\ k_{33} \\ k_{43} \end{bmatrix} = \begin{bmatrix} -\dfrac{12EI}{L^3} \\ -\dfrac{6EI}{L^2} \\ \dfrac{12EI}{L^3} \\ -\dfrac{6EI}{L^2} \end{bmatrix}$$

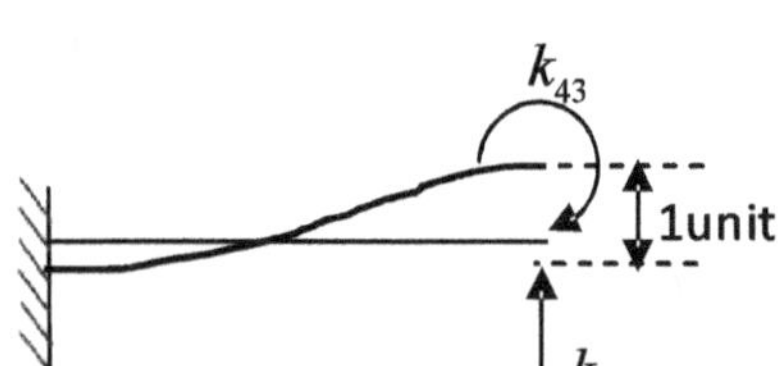

Figure 2.

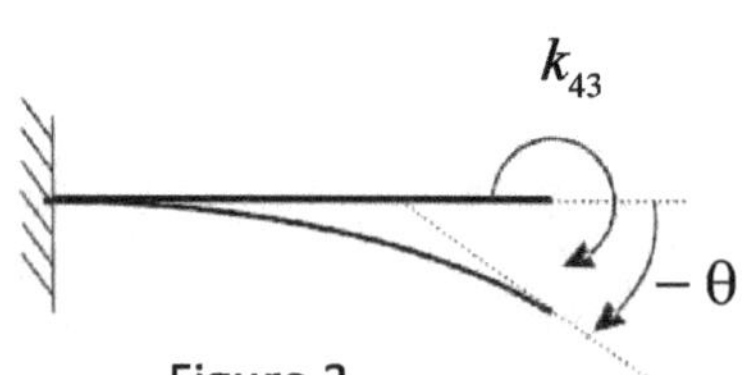

Figure 3.

Derivation of fourth column of stiffness matrix: $v_1 = \theta_1 = 0$, $v_2 = 0$, $\theta_2 = 1$, i.e., allow the fourth degree of freedom to occur and arrest all other DoF. (The deformed configuration is shown in Figure 2).

Initially you have a horizontal beam element. Since $v_1 = \theta_1 = 0$, we can fix node i. To produce a counterclockwise (positive) rotation or slope at node j (i.e., allowing fourth degree of freedom to occur), apply a counterclockwise moment k_{44}. $\theta_2 = \dfrac{k_{44} L}{EI}$. Refer Figure 1. This moment k_{44} will produce a upward deflection $\frac{k_{44}L^2}{2EI}$. This upward deflection should be canceled by applying a downward force k_{34} at node j. The downward deflection produced by k_{34} is $\frac{k_{34}L^3}{3EI}$. Refer Figure 2. Equating these two deflections $\frac{k_{44}L^2}{2EI}=\frac{k_{34}L^3}{3EI}$...(1) But this downward force k_{34} will also produce a negative slope at node j which is $\frac{k_{34}L^2}{2EI}$. Hence the rotation produced by k_{44} should be greater than that produced by k_{34} so that the resultant rotation is 1 radians. $\frac{k_{44}L}{EI}-\frac{k_{34}L^2}{2EI}=1$....(2) Refer Figure 3. Solving these two equations, k_{34} and k_{44} are found. The fixed end reaction force and the reaction moment are assumed to be acting upwards and counterclockwise, respectively. Now use force equilibrium equation to find fixed end reaction force k_{14} ... $\left(\sum F_y=0 \Rightarrow k_{14}+k_{34}=0\right)$ and moment equilibrium equation about node i to find fixed end reaction moment k_{24}.... $\left(\sum M_i=0 \Rightarrow k_{24}+k_{34}L+k_{44}=0\right)$.

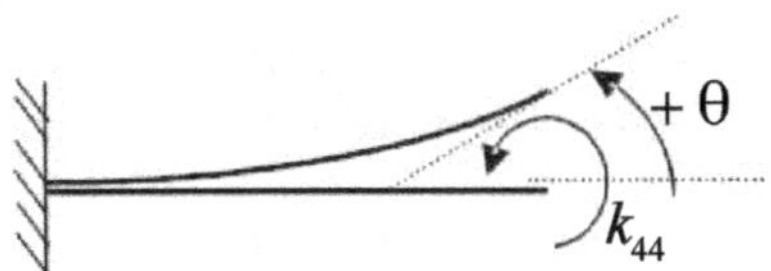

Figure 1.

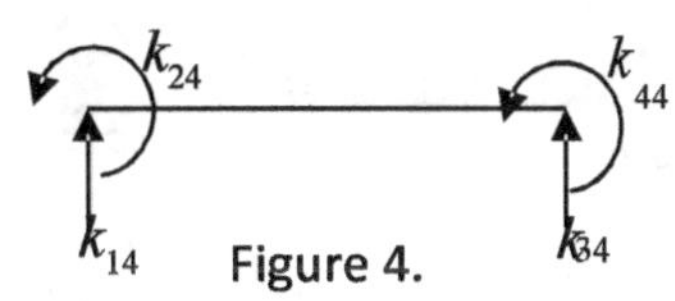

Figure 4.

$$\begin{bmatrix} k_{14} \\ k_{24} \\ k_{34} \\ k_{44} \end{bmatrix} = \begin{bmatrix} \frac{6EI}{L^2} \\ \frac{2EI}{L} \\ -\frac{6EI}{L^2} \\ \frac{4EI}{L} \end{bmatrix}$$

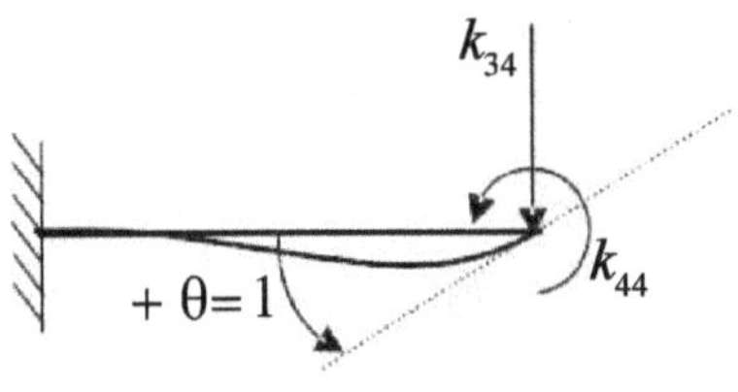

Figure 2.

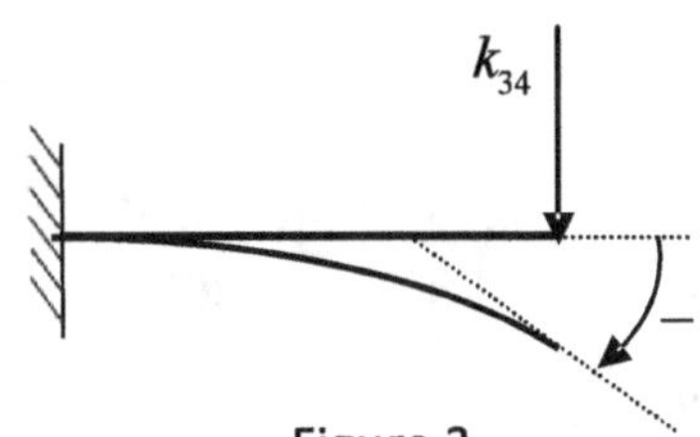

Figure 3.

Problem

Find the slopes at the supports and support reaction forces and support reaction moments for the beam shown in Figure. Take E=210 GPa, I = 2×10 $^{-4}$ m^4. Daryl Logan P4-24 page 208.

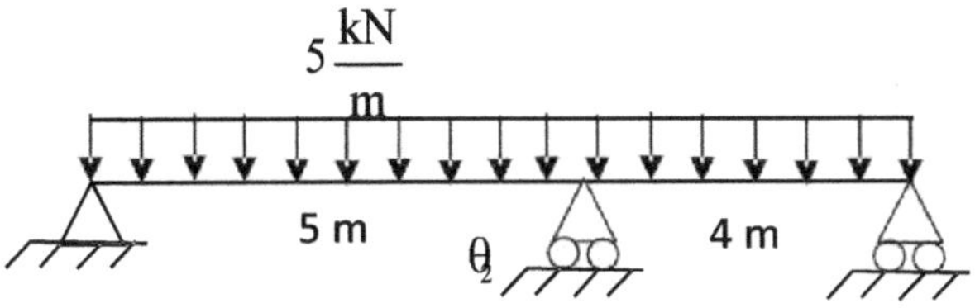

Finite element representation of the problem

v_2

Conversion of UDL into nodal forces and nodal moments

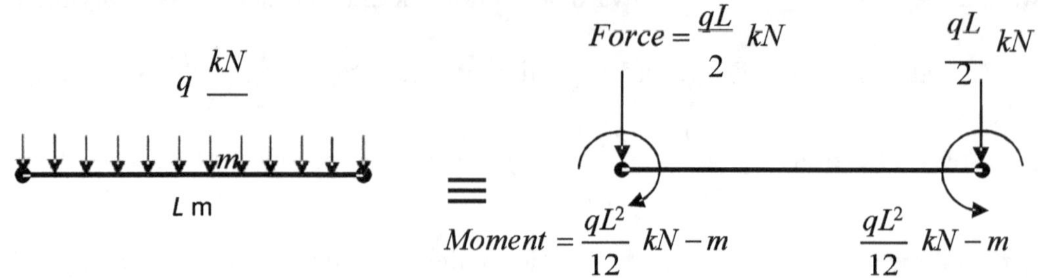

for element 1,

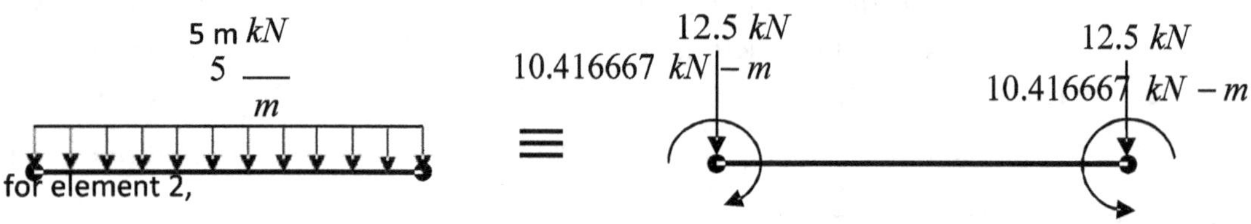

for element 2,

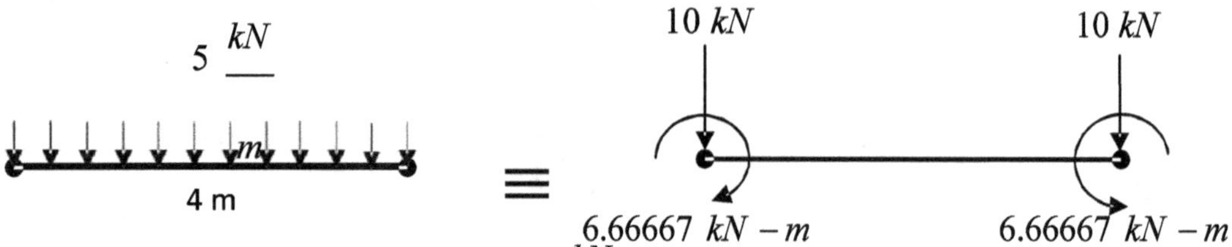

$$EI = 210 \text{ GPa} \times 2\times10^{-4} \text{ m}^4 = 210\times10^6 \frac{kN}{m^2} \times 2\times10^{-4} \text{ m}^4 = 42000 \text{ kN-m}^2$$

Stiffness matrix for element 1

$$[K]^{(1)} = \begin{bmatrix} 4,032 & 10,080 & -4032 & 10,080 \\ 10,080 & 33,600 & -10,080 & 16,800 \\ -4032 & -10,080 & 4,032 & -10,080 \\ 10,080 & 16,800 & -10,080 & 33,600 \end{bmatrix}$$

Stiffness matrix for element 2

$$[K]^{(2)} = \begin{bmatrix} 7,875 & 15,750 & -7,875 & 15,750 \\ 15,750 & 42,000 & -15,750 & 21,000 \\ -7,875 & -15,750 & 7,875 & -15,750 \\ 15,750 & 21,000 & -15,750 & 42,000 \end{bmatrix}$$

Assembly of finite element equations

Support reaction moments at all simply supported ends are zero. $M_1 = M_2 = M_3 = 0$

All support reaction forces are unknowns.

$$\begin{Bmatrix} F_1 \\ M_1 \\ F_2 \\ M_2 \\ F_2 \\ M_3 \end{Bmatrix} + \begin{Bmatrix} -12.5 \\ -10.416667 \\ -12.5-10 \\ 10.416667-6.66667 \\ -10 \\ 6.66667 \end{Bmatrix} = \begin{Bmatrix} F_1-12.5 \\ 0-10.416667 \\ F_2-22.5 \\ 0+3.75 \\ F_2-10 \\ 0+6.66667 \end{Bmatrix}$$

support reactions   applied forces

$v_1 = v_2 = v_3 = 0 \qquad \theta_1 = ? \qquad \theta_2 = ? \qquad \theta_3 = ?$

$$\begin{Bmatrix} F_1-12.5 \\ 0-10.416667 \\ F_2-22.5 \\ 0+3.75 \\ F_3-10 \\ 0+6.66667 \end{Bmatrix} = \begin{bmatrix} 4,032 & 10,080 & -4,032 & 10,080 & 0 & 0 \\ 10,080 & 33,600 & -10,080 & 16,800 & 0 & 0 \\ -4,032 & -10,080 & 4,032+7,875 & -10,080+15,750 & -7,875 & 15,750 \\ 10,080 & 16,800 & -10,080+15,750 & 33,600+42,000 & -15,750 & 21,000 \\ 0 & 0 & -7,875 & -15,750 & 7,875 & -15,750 \\ 0 & 0 & 15,750 & 21,000 & -15,750 & 42,000 \end{bmatrix} \begin{Bmatrix} 0 \\ \theta_1 \\ 0 \\ \theta_2 \\ 0 \\ \theta_3 \end{Bmatrix}$$

Eliminating the first, third and fifth rows and columns of the stiffness matrix, the reduced matrix becomes

$$\begin{Bmatrix} -10.416667 \\ 3.75 \\ 6.66667 \end{Bmatrix} = \begin{bmatrix} 33,600 & 16,800 & 0 \\ 16,800 & 75,600 & 21,000 \\ 0 & 21,000 & 42,000 \end{bmatrix} \begin{Bmatrix} \theta_1 \\ \theta_2 \\ \theta_3 \end{Bmatrix}$$

Solving these equations

$\theta_1 = -3.59623\times10^{-4}$ rad $\theta_2 = 9.9206349\times10^{-5}$ rad $\theta_3 = 1.0912698\times10^{-4}$ rad

Substituting these values in the assembled matrix to find the support reactions, we find

$$F_1 - 12.5 = \{10,080\times(-3.59623\times10^{-4})\} + \{10,080\times9.92\times10^{-5}\}$$

$$F_2 - 22.5 = \{10,080\times(-3.59623\times10^{-4})\} + \{5,670\times9.92\times10^{-5}\} + \{15,750\times1.0913\times10^{-4}\}$$

$$F_3 - 10 = \{15,750\times9.92\times10^{-5}\} - \{15,750\times1.0913\times10^{-4}\}$$

$F_1 - 12.5 = -2.625$ kN $F_1 = 9.875$ kN

$F_2 - 22.5 = 5.9062$ kN which means $F_2 = 28.406$ kN

$F_3 - 12.5 = -3.2812$ kN $F_3 = 6.71869$ kN

It is verified that the total applied load $\left(5\dfrac{kN}{m}\times 5m = 45kN\right)$ is equal to the sum of the support reaction forces (9.875+28.406+6.71869 = 45 kN).

Total force and moment diagram

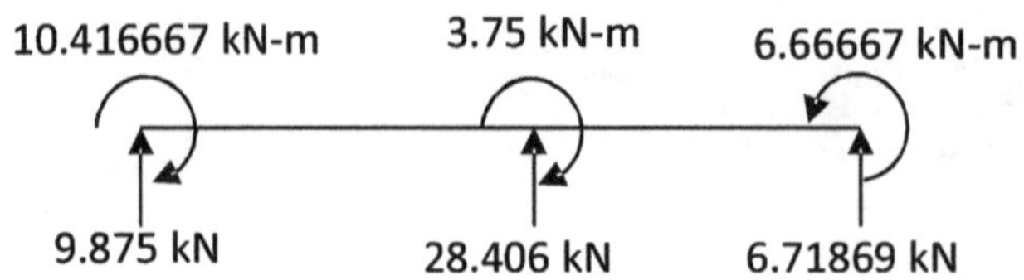

Individual force and moment diagrams

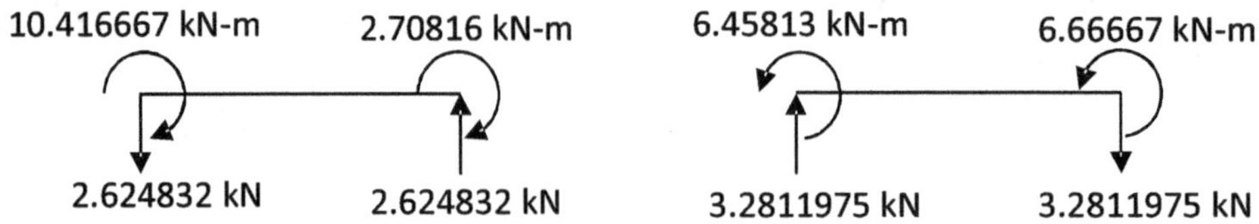

Individual force and moment calculations

Element 1

$$F_1^{(1)} = \{10,080\times(-3.59623\times10^{-4})\} + \{10,080\times9.92\times10^{-5}\} = -2.624832 \text{ kN}$$

$$M_1^{(1)} = \{33,600\times(-3.59623\times10^{-4})\} + \{16,800\times9.92\times10^{-5}\} = -10.416 \text{ kN-m}$$

$$F_2^{(1)} = -\{10,080\times(-3.59623\times10^{-4})\} - \{10,080\times9.92\times10^{-5}\} = 2.624832 \text{ kN}$$

$$M_2^{(1)} = \{16,800\times(-3.59623\times10^{-4})\} + \{33,60\times9.92\times10^{-5}\} = -2.70816 \text{ kN-m}$$

Element 2

$$F_1^{(1)} = \{15,750\times9.92\times10^{-5}\} + \{15,750\times1.0913\times10^{-4}\} = 3.2811975 \text{ kN}$$

$$M_1^{(1)} = \{42,000\times9.92\times10^{-5}\} + \{21,000\times1.0913\times10^{-4}\} = 6.45813 \text{ kN-m}$$

$$F_2^{(1)} = \{15,750\times9.92\times10^{-5}\} - \{15,750\times1.0913\times10^{-4}\} = -3.2811975 \text{ kN}$$

$$M_2^{(1)} = \{21,000\times9.92\times10^{-5}\} + \{42,000\times1.0913\times10^{-4}\} = 6.66667 \text{ kN-m}$$

PROBLEM

Given that E=210 GPa and I=4×10 $^{-4}$ m^4, cross section of the beam is constant. Determine the deflection and slope at point C. calculate the reaction forces and moments. DARYL LOGAN P 171-172

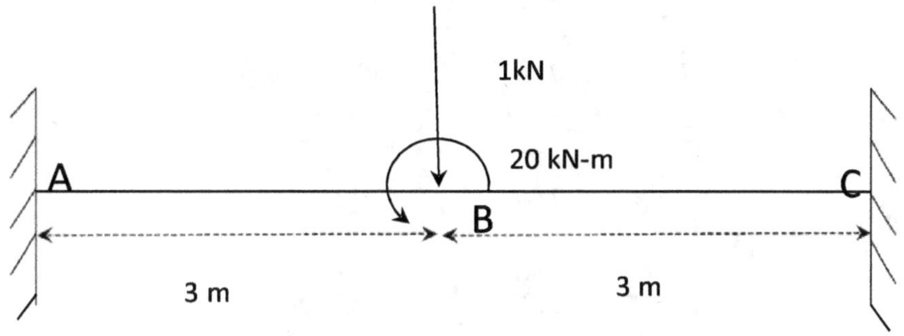

Solution:-

Degree of freedom in numbers:-

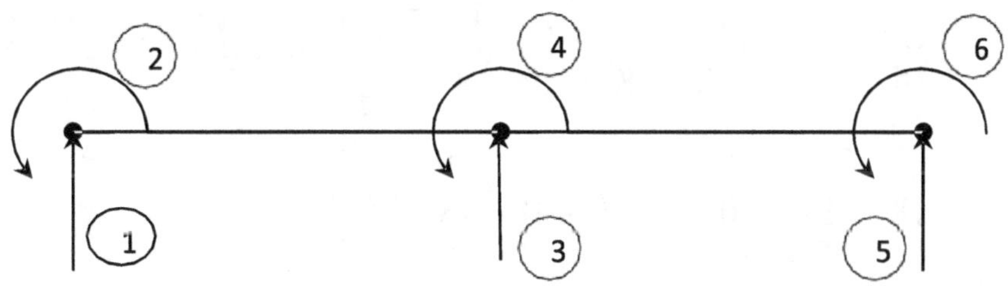

Degree of freedom of forces and moments:-

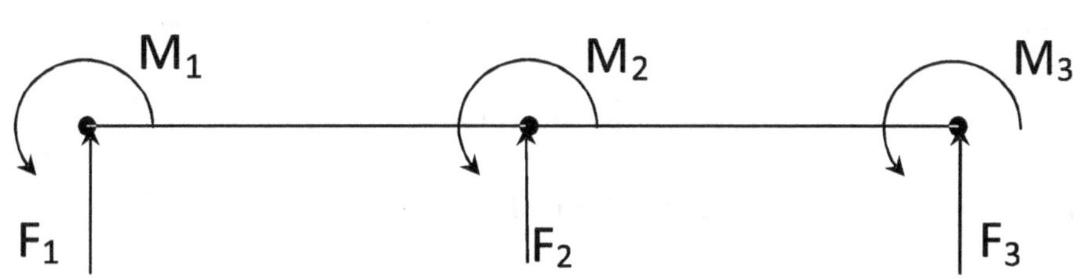

Degree of freedom of displacement and rotation:-

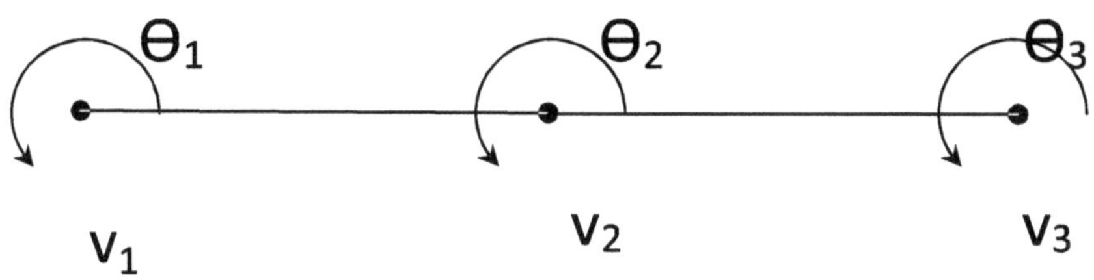

Stiffness matrix for element 1 and 2:-

$$[K]^1 = [K]^2 = \begin{bmatrix} \frac{12EI}{l^3} & \frac{6EI}{l^2} & -\frac{12EI}{l^3} & \frac{6EI}{l^2} \\ \frac{6EI}{l^2} & \frac{4EI}{l} & -\frac{6EI}{l^2} & \frac{2EI}{l} \\ -\frac{12EI}{l^3} & -\frac{6EI}{l^2} & \frac{12EI}{l^3} & -\frac{6EI}{l^2} \\ \frac{6EI}{l^2} & \frac{2EI}{l} & -\frac{6EI}{l^2} & \frac{4EI}{l} \end{bmatrix}$$

$$[K]^1 = [K]^2 = 3.1\times10^6 \begin{bmatrix} 12 & 18 & -12 & 18 \\ 18 & 36 & -18 & 18 \\ -12 & -18 & 12 & -18 \\ 18 & 18 & -18 & 36 \end{bmatrix}$$

Assembling:-

$$\begin{Bmatrix} F_1 \\ M_1 \\ F_2 \\ M_2 \\ F_3 \\ M_3 \end{Bmatrix} = 3.1\times10^6 \begin{bmatrix} 12 & 18 & -12 & 18 & 0 & 0 \\ 18 & 36 & -18 & 18 & 0 & 0 \\ -12 & -18 & 24 & 0 & 0 & 0 \\ 18 & 18 & 0 & 72 & -18 & 18 \\ 0 & 0 & -12 & -18 & 12 & -18 \\ 0 & 0 & 18 & 18 & -18 & 36 \end{bmatrix} \begin{Bmatrix} v_1 \\ \theta_1 \\ v_2 \\ \theta_2 \\ v_3 \\ \theta_3 \end{Bmatrix}$$

Boundary condition:-

F_2=-10 kN; M_2=20 kN-m $\qquad v_1 = v_3 = \Theta_1 = \Theta_3 = 0$

Therefore first, second, fifth, sixth columns are ineffective

and hence the reduced matrix is given by

$$\begin{Bmatrix} F_2 \\ m_2 \end{Bmatrix} = 3.1\times10^6 \begin{bmatrix} 24 & 0 \\ 0 & 72 \end{bmatrix} \begin{Bmatrix} v_2 \\ \theta_2 \end{Bmatrix}$$

Deflection and slope at point c:-

V_2= -1.34×10^{-4} m = -0.134 mm

Θ_2= 8.96×10^{-5} rad

Reaction forces and moments:-

$$\begin{Bmatrix} F_2 \\ m_2 \\ F_3 \\ m_3 \end{Bmatrix} = 3.1\times10^6 \begin{bmatrix} -12 & 18 \\ -18 & 18 \\ -12 & -18 \\ 18 & 18 \end{bmatrix}$$

F_1=10000N

M_1=12500N-m

F_3=0

M_3= -2500N-m

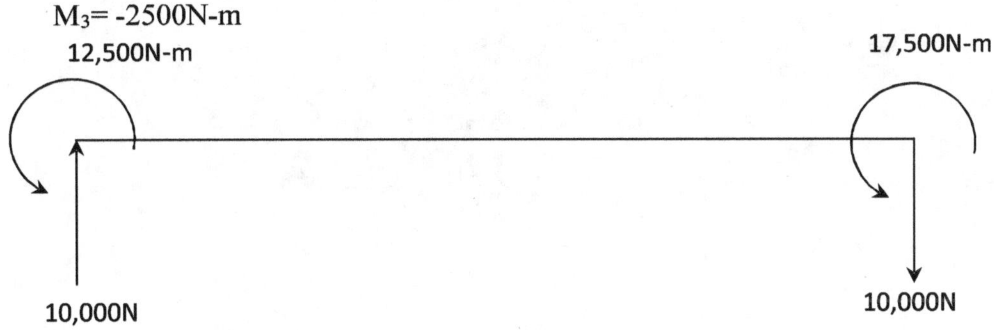

individual element forces and moments are

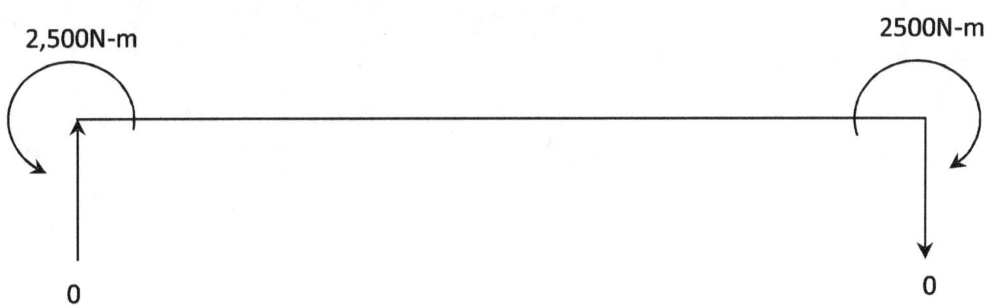

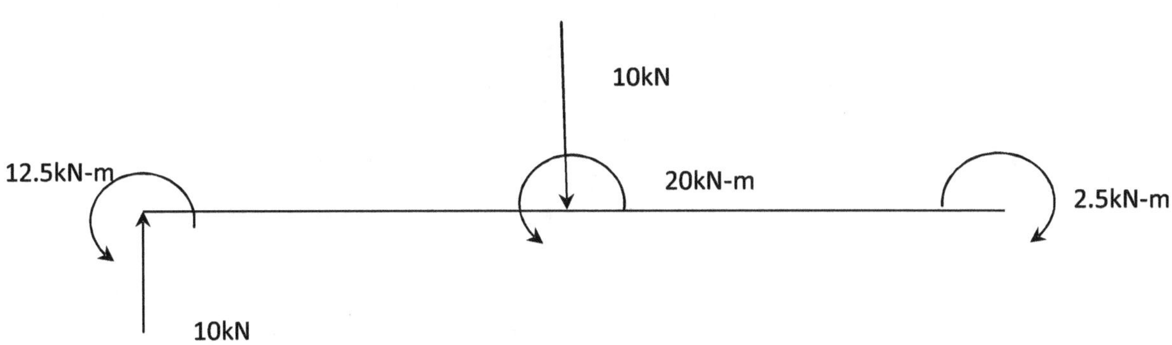

UNIT III

TWO DIMENSIONAL FINITE ELEMENT ANALYSIS

3.1 INTRODUCTION

Two dimensional elements are defined by three or more nodes in a two dimensional plane (i.e., x, y plane). The basic element useful for two dimensional analysis is the triangular element.

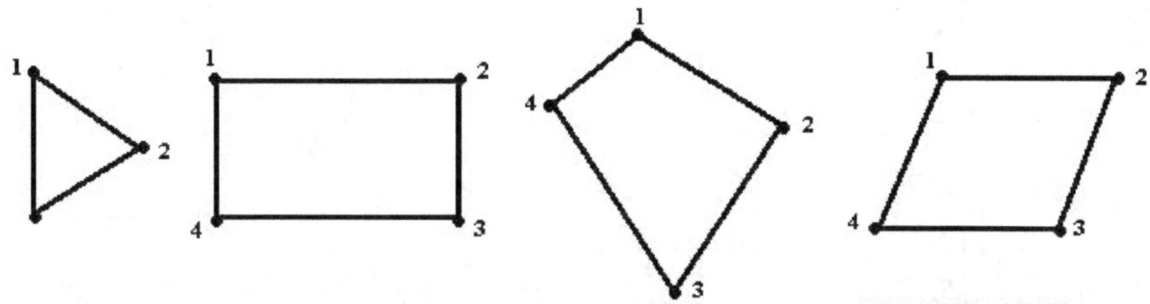

TRIANGLE RECTANGLE QUADRILATERAL PARALLELOGRAM

➢ **Plane Stress and Plane Strain**

The 2d element is extremely important for the Plane Stress analysis and Plane Strain analysis.

Plane Stress Analysis:

It is defined to be a state of stress in which the normal stress (σ) and shear stress (τ) directed perpendicular to the plane are assumed to be zero.

Plane Strain Analysis:

It is defined to be a state of strain in which the normal to the xy plane and the shear strain are assumed to be zero.

3.2 THREE NODED LINEAR TRIANGULAR ELEMENT

The physical domain considered is geometrically a 2-Dimensional domain, i.e., an area with uniform thickness and the single variable can be one of pressure, temperature, etc. (a scalar quantity, not a vector quantity). An example is the temperature distribution in a plate. At each point there can be only one temperature. We consider such an area meshed with triangular elements. Each triangular element has three nodes, (i.e., one node at each corner). Let us consider one such element with coordinates $(x_1, y_1), (x_2, y_2)$ and (x_3, y_3). The single variable (for example, temperature) at these nodes 1, 2 and 3 are $u_1 u_2$ and u_3, respectively. If so, then the unknown single variable u (temperature) at any non-nodal point (x,y) in the 2-D domain can be expressed in terms of the known nodal variables (temperatures) $u_1 u_2$ and u_3.

Let us assume that the single variable can be expressed as

$$u = c_1 + c_2 x + c_3 y$$

In order to find the three unknowns c_1, c_2 and c_3, we apply the boundary conditions

at (x_1, y_1), $u = u_1 \Rightarrow u = c_1 + c_2 x_1 + c_3 y_1$

at (x_2, y_2), $u = u_2 \Rightarrow u = c_1 + c_2 x_2 + c_3 y_2$

at (x_3, y_3), $u = u_3 \Rightarrow u = c_1 + c_2 x_3 + c_3 y_3$

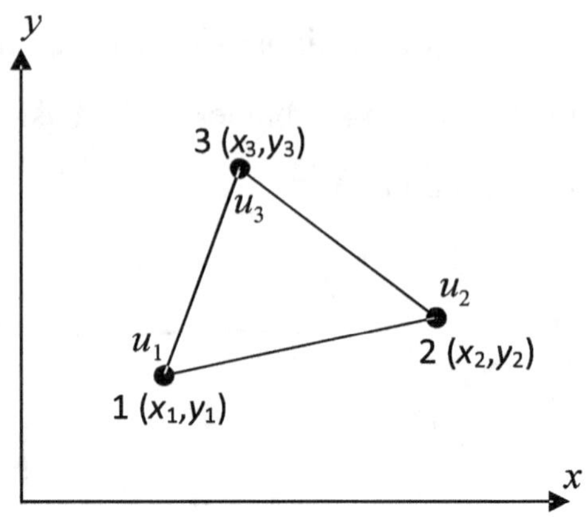

Writing the above three equations in matrix form

$$\begin{Bmatrix} u_1 \\ u_2 \\ u_3 \end{Bmatrix} = \begin{bmatrix} 1 & x_1 & y_1 \\ 1 & x_2 & y_2 \\ 1 & x_3 & y_3 \end{bmatrix} \begin{Bmatrix} c_1 \\ c_2 \\ c_3 \end{Bmatrix}$$

We need to find c_1, c_2 and c_3

$$\begin{Bmatrix} c_1 \\ c_2 \\ c_3 \end{Bmatrix} = \begin{bmatrix} 1 & x_1 & y_1 \\ 1 & x_2 & y_2 \\ 1 & x_3 & y_3 \end{bmatrix}^{-1} \begin{Bmatrix} u_1 \\ u_2 \\ u_3 \end{Bmatrix}$$

$$\begin{bmatrix} 1 & x_1 & y_1 \\ 1 & x_2 & y_2 \\ 1 & x_3 & y_3 \end{bmatrix}^{-1} = \frac{1}{2A} \begin{bmatrix} \alpha_1 & \alpha_2 & \alpha_3 \\ \beta_1 & \beta_2 & \beta_3 \\ \gamma_1 & \gamma_2 & \gamma_3 \end{bmatrix}$$ where $2A = \alpha_1 + \alpha_2 + \alpha_3$ and

$\alpha_i = x_j y_k - x_k y_j$ $\qquad \beta_i = y_j - y_k$ $\qquad \gamma_i = -(x_j - x_k)$

$\alpha_1 = x_2 y_3 - x_3 y_2$ $\qquad \beta_1 = y_2 - y_3$ $\qquad \gamma_1 = -(x_2 - x_3)$

$\alpha_2 = x_3 y_1 - x_1 y_3$ $\qquad \beta_2 = y_3 - y_1$ $\qquad \gamma_2 = -(x_3 - x_1)$

$\alpha_3 = x_1 y_2 - x_2 y_1$ $\qquad \beta_3 = y_1 - y_2$ $\qquad \gamma_3 = -(x_1 - x_2)$

A is the area of the triangle.

$$\begin{Bmatrix} c_1 \\ c_2 \\ c_3 \end{Bmatrix} = \frac{1}{2A} \begin{bmatrix} \alpha_1 & \alpha_2 & \alpha_3 \\ \beta_1 & \beta_2 & \beta_3 \\ \gamma_1 & \gamma_2 & \gamma_3 \end{bmatrix} \begin{Bmatrix} u_1 \\ u_2 \\ u_3 \end{Bmatrix}$$

Substituting the values of c_1, c_2 and c_3 in $u = c_1 + c_2 x + c_3 y$, we get

$$u = N_1 u_1 + N_2 u_2 + N_3 u_3$$

where $N_i = \frac{1}{2A}(\alpha_i + \beta_i x + \gamma_i y)$, $\qquad i = 1, 2, 3$

3.3 FOUR NODED LINEAR RECTANGULAR ELEMENT

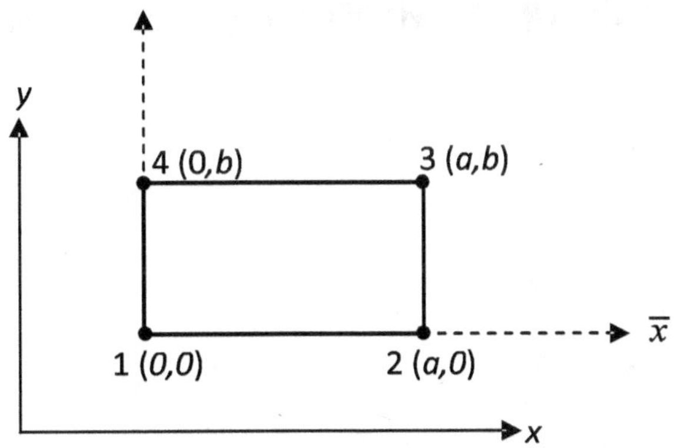

Let us assume that the single variable can be expressed as

$$u(\bar{x},\bar{y}) = c_1 + c_2\bar{x} + c_3\bar{y} + c_4\bar{x}\bar{y} \quad \ldots\ldots\ldots\ldots\ldots\ldots(1)$$

This polynomial contains four linearly independent terms and is linear in x and y with a bilinear term in x and y. The polynomial requires an element with four nodes. There are two possible geometric shapes: a triangle with the fourth node at the centroid of the triangle or a rectangle with nodes at the vertices.

A triangle with a fourth node at the center does not provide a single-valued variation of u at inter-element boundaries, resulting in incompatible variation of u at inter-element boundaries and is therefore not admissible.

The linear rectangular element is a compatible element because on any side, the single variable u varies only linearly and there are two nodes to uniquely define it.

Here we consider an approximation of the form given in eqauation (1) and use a rectangular element with sides a and b. For the sake of convenience we choose a local coordinate system $(\bar{x},\bar{y})$ to derive the interpolation functions.

In order to find the three unknowns c_1, c_2 and c_3, we apply the boundary conditions

at $(0, 0)$, $u = u_1$ $\quad \Rightarrow = u \quad c_1$

at $(a, 0)$, $u = u_2$ $\quad \Rightarrow = u \quad c_1 + c_2 a$

at (a,b), $u = u_3$ $\quad \Rightarrow = u \quad c_1 + c_2 a + c_3 ab$

at $(0,b)$, $u = u_3$ $\quad \Rightarrow = u \quad c_1 + c_4 b$

Solving for c_1, c_2, c_3 and c_4

y

$$c_1 = u_1, \quad c_1 = \frac{u_2 - u_1}{a}, \quad c_3 = \frac{u_4 - u_1}{b}, \quad c_4 = \frac{u_1 - u_2 + u_3 - u_4}{ab},$$

3.4 TWO-VARIABLE 3-NODED LINEAR TRIANGULAR ELEMENT

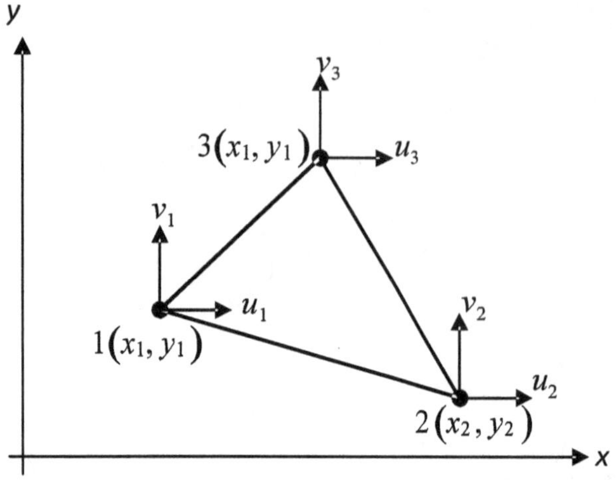

Figure shows a 2-D two-variable linear triangular element with three nodes and the two dof at each node. The nodes are placed at the corners of the triangle. The two variables (dof) are displacement in *x*-direction (u) and displacement in *y*-direction (v). Since each node has two dof, a single element has 6 dof. The nodal displacement vector is given by

$$\{U\} = \begin{Bmatrix} u_1 \\ v_1 \\ u_2 \\ v_2 \\ u_3 \\ v_3 \end{Bmatrix}$$

We select a linear displacement function for each dof as

$$u(x,y) = c_1 + c_2 x + c_3 y$$
$$v(x,y) = c_4 + c_5 x + c_6 y$$

where $u(x,y)$ and $v(x,y)$ describe displacements at any interior point (x,y) of the element.

The above two algebraic equations can also be written as

$$\begin{Bmatrix} u \\ v \end{Bmatrix} = \begin{bmatrix} 1 & x & y & 0 & 0 & 0 \\ 0 & 0 & 0 & 1 & x & y \end{bmatrix} \begin{Bmatrix} c_1 \\ c_2 \\ c_3 \\ c_4 \\ c_5 \\ c_6 \end{Bmatrix}$$

Using steps we had developed for the 2-D single-variable linear triangular element, we can write

$$\begin{Bmatrix} c_1 \\ c_2 \\ c_3 \end{Bmatrix} = \frac{1}{2A}\begin{bmatrix} \alpha_1 & \alpha_2 & \alpha_3 \\ \beta_1 & \beta_2 & \beta_3 \\ \gamma_1 & \gamma_2 & \gamma_3 \end{bmatrix}\begin{Bmatrix} u_1 \\ u_2 \\ u_3 \end{Bmatrix}$$

$$\begin{Bmatrix} c_4 \\ c_5 \\ c_6 \end{Bmatrix} = \frac{1}{2A}\begin{bmatrix} \alpha_1 & \alpha_2 & \alpha_3 \\ \beta_1 & \beta_2 & \beta_3 \\ \gamma_1 & \gamma_2 & \gamma_3 \end{bmatrix}\begin{Bmatrix} v_1 \\ v_2 \\ v_3 \end{Bmatrix}$$

and using the interpolation functions we had developed for the 2-D single-variable linear triangularelement, we can write

$$u(x,y) = N_1 u_1 + N_2 u_2 + N_3 u_3$$
$$v(x,y) = N_1 v_1 + N_2 v_2 + N_3 v_3$$

where

$$N_i = \frac{1}{2A}(\alpha_i + \beta_i x + \gamma_i y), \qquad i = 1, 2, 3$$

Writing the above equations in matrix form $\begin{Bmatrix} u(x,y) \\ v(x,y) \end{Bmatrix} = \begin{bmatrix} N_1 & 0 & N_2 & 0 & N_3 & 0 \\ 0 & N_1 & 0 & N_2 & 0 & N_3 \end{bmatrix}\begin{Bmatrix} u_1 \\ v_1 \\ u_2 \\ v_2 \\ u_3 \\ v_3 \end{Bmatrix}$ The

$$\{U\} = [N]\{a\}$$

strains associated with the two-dimensional element are given by

$$\{\varepsilon\}=\begin{Bmatrix}\varepsilon_x\\ \varepsilon_y\\ \gamma_{xy}\end{Bmatrix}=\begin{Bmatrix}\dfrac{\partial u}{\partial x}\\ \dfrac{\partial v}{\partial y}\\ \dfrac{\partial u}{\partial y}+\dfrac{\partial v}{\partial x}\end{Bmatrix}$$

and note that

$$\frac{\partial N_i}{\partial x}=\beta_i \qquad \frac{\partial N_i}{\partial y}=\gamma_i$$

$$\frac{\partial u}{\partial x}=\frac{\partial}{\partial x}(N_1u_1+N_2u_2+N_3u_3)=\left\langle\frac{\partial N_1}{\partial x}\quad\frac{\partial N_2}{\partial x}\quad\frac{\partial N_3}{\partial x}\right\rangle\begin{Bmatrix}u_1\\ u_2\\ u_3\end{Bmatrix}=\left\langle\frac{\partial N_1}{\partial x}\quad 0\quad\frac{\partial N_2}{\partial x}\quad 0\quad\frac{\partial N_3}{\partial x}\quad 0\right\rangle\begin{Bmatrix}u_1\\ v_1\\ u_2\\ v_2\\ u_3\\ v_3\end{Bmatrix}$$

$$\frac{\partial v}{\partial y}=\frac{\partial}{\partial y}(N_1v_1+N_2v_2+N_3v_3)=\left\langle\frac{\partial N_1}{\partial y}\quad\frac{\partial N_2}{\partial y}\quad\frac{\partial N_3}{\partial y}\right\rangle\begin{Bmatrix}v_1\\ v_2\\ v_3\end{Bmatrix}=\left\langle 0\quad\frac{\partial N_1}{\partial y}\quad 0\quad\frac{\partial N_2}{\partial y}\quad 0\quad\frac{\partial N_3}{\partial y}\right\rangle\begin{Bmatrix}u_1\\ v_1\\ u_2\\ v_2\\ u_3\\ v_3\end{Bmatrix}$$

$$\frac{\partial u}{\partial y}+\frac{\partial v}{\partial x}=\frac{\partial}{\partial y}(N_1u_1+N_2u_2+N_3u_3)+\frac{\partial}{\partial x}(N_1v_1+N_2v_2+N_3v_3)=\left\langle\frac{\partial N_1}{\partial y}\quad\frac{\partial N_1}{\partial x}\quad\frac{\partial N_2}{\partial y}\quad\frac{\partial N_2}{\partial x}\quad\frac{\partial N_3}{\partial y}\quad\frac{\partial N_3}{\partial x}\right\rangle\begin{Bmatrix}u_1\\ v_1\\ u_2\\ v_2\\ u_3\\ v_3\end{Bmatrix}$$

$$\{\varepsilon\}=\begin{Bmatrix}\varepsilon_x\\ \varepsilon_y\\ \gamma_{xy}\end{Bmatrix}=\begin{bmatrix}\dfrac{\partial N_1}{\partial x} & 0 & \dfrac{\partial N_2}{\partial x} & 0 & \dfrac{\partial N_3}{\partial x} & 0\\ 0 & \dfrac{\partial N_1}{\partial y} & 0 & \dfrac{\partial N_2}{\partial y} & 0 & \dfrac{\partial N_3}{\partial y}\\ \dfrac{\partial N_1}{\partial y} & \dfrac{\partial N_1}{\partial x} & \dfrac{\partial N_2}{\partial y} & \dfrac{\partial N_2}{\partial x} & \dfrac{\partial N_3}{\partial y} & \dfrac{\partial N_3}{\partial x}\end{bmatrix}\begin{Bmatrix}u_1\\ v_1\\ u_2\\ v_2\\ u_3\\ v_3\end{Bmatrix}=\begin{bmatrix}\beta_1 & 0 & \beta_2 & 0 & \beta_3 & 0\\ 0 & \gamma_1 & 0 & \gamma_2 & 0 & \gamma_3\\ \gamma_1 & \beta_1 & \gamma_2 & \beta_2 & \gamma_3 & \beta_3\end{bmatrix}\begin{Bmatrix}u_1\\ v_1\\ u_2\\ v_2\\ u_3\\ v_3\end{Bmatrix}$$

$$\{\varepsilon\}_{3\times1}=[B]_{3\times6}\{a\}_{6\times1}$$

$$\{\sigma\} = [D]\{\varepsilon\}$$

$$\begin{Bmatrix} \sigma_x \\ \sigma_y \\ \tau_{xy} \end{Bmatrix} = [D] \begin{Bmatrix} \varepsilon_x \\ \varepsilon_y \\ \gamma_{xy} \end{Bmatrix}$$

$$\{\sigma\} = [D][B]\{a\}$$

The stiffness matrix is given by $\underset{6\times6}{[K]} = \int_{Volume} \underset{6\times3}{[B]^T} \underset{3\times3}{[D]} \underset{3\times6}{[B]} \, dV = t \iint_{Area} [B]^T [D][B] \, dx dy$.

where t is the thickness of the plate. The integrand $[B]^T [D][B]$ is not a function of x and y and hence can be taken outside the integral to yield

$$[K] = tA[B]^T [D][B]$$

D matrix is the material constitutive matrix, either for the plane-stress case or for the plane-strain case depending on the problem in hand.

and substituting them back in u, we get

$$u = N_1 u_1 + N_2 u_2 + N_3 u_3 + N_4 u_4$$

$$\text{where } N_1 = \left(1 - \frac{x}{a}\right)\left(1 - \frac{y}{b}\right)$$

$$N_2 = \frac{x}{a}\left(1 - \frac{y}{b}\right)$$

$$N_3 = \frac{x}{a}\frac{y}{b}$$

$$N_4 = \left(1 - \frac{x}{a}\right)\frac{y}{b}$$

3.5 STRAIN – STRESS RELATION

$$\varepsilon_x = \frac{\sigma_x}{E} - \nu\frac{\sigma_y}{E} - \nu\frac{\sigma_z}{E}$$

$$\varepsilon_y = -\nu\frac{\sigma_x}{E} + \frac{\sigma_y}{E} - \nu\frac{\sigma_z}{E}$$

$$\varepsilon = -\nu\frac{\sigma_x}{E} - \nu\frac{\sigma_y}{E} + \frac{\sigma_z}{E}$$

3.5.1 Plane stress conditions

σ_x, σ_y and τ_{xy} are present.

$\sigma_z = \tau_{xz} = \tau_{yz} = 0.$

since $\sigma_z = 0$, from equations1 and 2

$$\varepsilon_x = \frac{\sigma_x}{E} - \nu\frac{\sigma_y}{E}$$

$$\varepsilon_y = -\nu\frac{\sigma_x}{E} + \frac{\sigma_y}{E}$$

solving the above two equations

for σ_x and σ_y, we get

$$\sigma_x = \frac{E}{(1-\nu^2)}\left(\varepsilon_x + \nu\varepsilon_y\right) \text{and}$$

$$\sigma_y = \frac{E}{(1-\nu^2)}\left(\nu\varepsilon_x + \varepsilon_y\right)$$

$$\tau_{xy} = G\,\gamma_{xy} = \frac{E}{2(1+\nu)}\gamma_{xy}$$

$$= \frac{(1-\nu^2)\,E}{(1-\nu^2)\,2(1+\nu)}\gamma_{xy}$$

$$= \frac{(1+\nu)(1-\nu)\quad E}{(1-\nu^2)\quad 2(1+\nu)}\gamma_{xy}$$

$$\tau_{xy} = \frac{E\quad(1-\nu)}{(1-\nu^2)\quad 2}\gamma_{xy}$$

writing σ_x, σ_y and τ_{xy} in a matrix form

$$\begin{Bmatrix}\sigma_x\\ \sigma_y\\ \tau_{xy}\end{Bmatrix} = \frac{E}{(1-\nu^2)}\begin{bmatrix}1 & \nu & 0\\ \nu & 1 & 0\\ 0 & 0 & \frac{(1-\nu)}{2}\end{bmatrix}\begin{Bmatrix}\varepsilon_x\\ \varepsilon_y\\ \gamma_{xy}\end{Bmatrix}$$

3.5.2 Plane strain conditions

σ_x, σ_y and τ_{xy} are present.

$\varepsilon_z = \gamma_{xz} = \gamma_{yz} = 0.$

σ_z is not zero.

since $\varepsilon_z = 0$, we get from equation 3

$$\sigma_z = \nu(\sigma_x + \sigma_x)$$

substituting σ_z in equations1 and 2

$$\varepsilon_x = \frac{\sigma_x}{E} - \nu\frac{\sigma_y}{E} - \nu^2\frac{(\sigma_x + \sigma_y)}{E}$$

$$\varepsilon_y = -\nu\frac{\sigma_x}{E} + \frac{\sigma_y}{E} - \nu^2\frac{(\sigma_x + \sigma_y)}{E}$$

rearranging the terms we get

$$\varepsilon_x = \frac{\sigma_x}{E}(1-\nu^2) - \frac{\sigma_y}{E}\nu(1+\nu)$$

$$\varepsilon_y = -\frac{\sigma_x}{E}\nu(1+\nu) + \frac{\sigma_y}{E}(1-\nu^2)$$

mutiplying by X by $\nu$and Y by $(1-\nu)$

$$\nu\varepsilon_x = \frac{\sigma_x}{E}\nu(1-\nu^2) - \frac{\sigma_y}{E}\nu^2(1+\nu)$$

$$(1-\nu)\varepsilon_y = -\frac{\sigma_x}{E}\nu(1-\nu)(1+\nu) + \frac{\sigma_y}{E}(1-\nu)(1-\nu^2)$$

$$= -\frac{\sigma_x}{E}\nu(1-\nu^2) + \frac{\sigma_y}{E}(1-\nu)(1-\nu^2)$$

adding the above two equations to eliminate σ_x

$$\nu\varepsilon_x + (1-\nu)\varepsilon_y = \frac{\sigma_y}{E}\left[-\nu^2(1+\nu) + (1-\nu)(1-\nu^2)\right]$$

$$\nu\varepsilon_x + (1-\nu)\varepsilon_y = \frac{\sigma_y}{E}\left[-\nu^2(1+\nu) + (1-\nu)(1+\nu)(1-\nu)\right]$$

$$\nu\varepsilon_x + (1-\nu)\varepsilon_y = \frac{\sigma_y}{E}(1+\nu)\left[-\nu^2 + (1-\nu)(1-\nu)\right]$$

$$\frac{\nu\varepsilon_x + (1-\nu)\varepsilon_y}{(1+\nu)} = \frac{\sigma_y}{E}\left[-\nu^2 + 1 + \nu^2 - 2\nu\right]$$

$$\sigma_y = E\frac{\nu\varepsilon_x + (1-\nu)\varepsilon_y}{(1+\nu)(1-2\nu)}$$

similarly

$$\sigma_x = E\frac{(1-\nu)\varepsilon_x + \nu\varepsilon_y}{(1+\nu)(1-2\nu)}$$

and as before

$$\tau_{xy} = \frac{E}{2(1+\nu)}\gamma_{xy}$$

writing σ_x, σ_y and τ_{xy} in matrix form

$$\begin{Bmatrix} \sigma_x \\ \sigma_y \\ \tau_{xy} \end{Bmatrix} = \frac{E}{(1+\nu)(1-2\nu)} \begin{bmatrix} (1-\nu) & \nu & 0 \\ \nu & (1-\nu) & 0 \\ 0 & 0 & \frac{(1-2\nu)}{2} \end{bmatrix} \begin{Bmatrix} \varepsilon_x \\ \varepsilon_x \\ \gamma_{xy} \end{Bmatrix}$$

It is difficult to represent the curved boundaries by straight edges element a large number of element may be used to obtain reasonable resembalance between original body and the assemblage

Two-Dimensional Problems

Review of the Basic Theory

In general, the stresses and strains in a structure consist of six components:

$s_x, s_y, s_z, t_{xy}, t_{yz}, t_{zx}$ for stresses,

and $e_x, e_y, e_z, g_{xy}, g_{yz}, g_{zx},$ for strains.

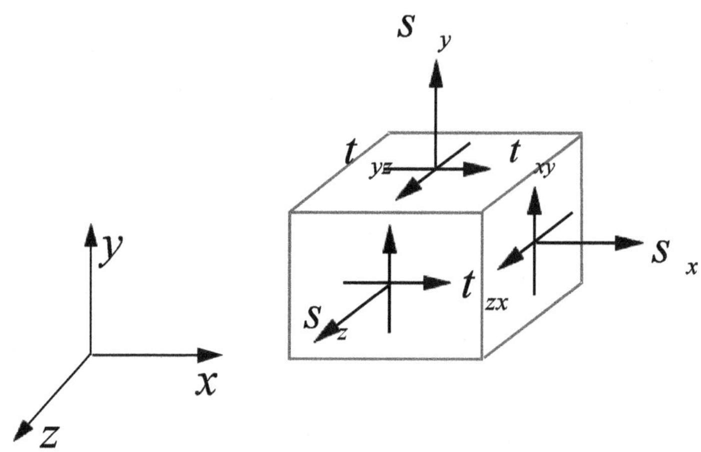

Under contain conditions, the state of stresses and strains can be simplified. A general 3-D structure analysis can, therefore, be reduced to a 2-D analysis.

$$t_{zx} \quad 0 \qquad (e_z \quad 0) \qquad (1)$$

Plane (2-D) Problems

Plane stress:

$s_z \quad t_{yz}$

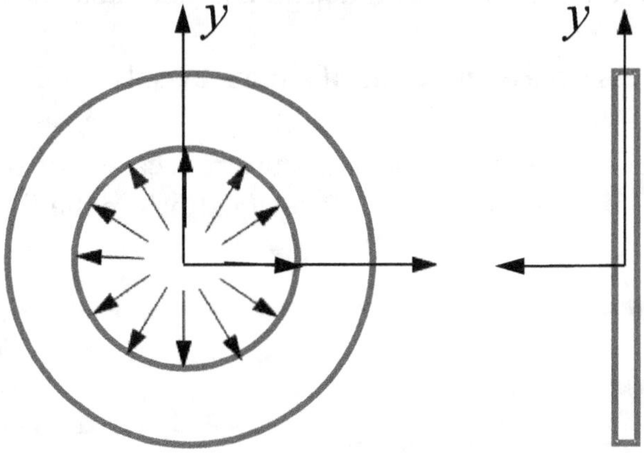

Plane strain:

A long structure with a uniform cross section and transverse loading along its length (z-direction).

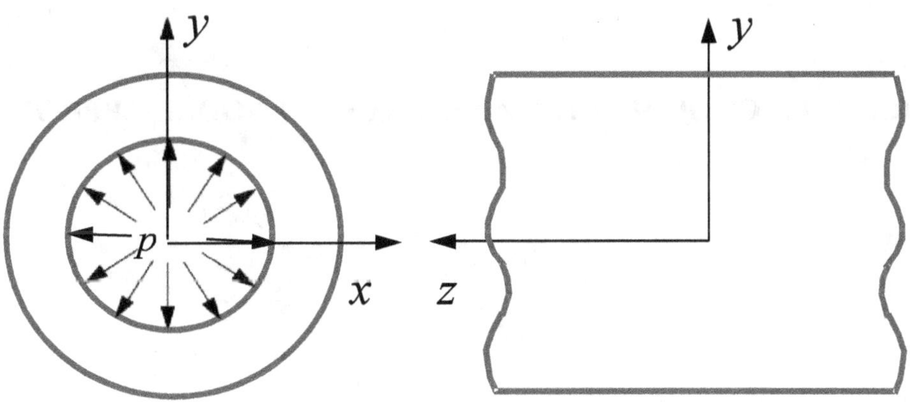

Stress-Strain-Temperature (Constitutive) Relations

For elastic and isotropic materials, we have,

$$\begin{array}{cccccc} e_x & 1/E & n/E & 0 & s_x & e_x \; 0 \\ e_y & n/E & 1/E & 0 & s_y & e_y \; 0 \\ g_{xy} & 0 & 0 & 1/G & t_{xy} & g_{xy} \; 0 \end{array}$$

where e_0 is the initial strain, E the Young's modulus, n the Poisson's ratio and G the shear modulus. Note that, $G = \frac{E}{2(1-n)}$

which means that there are only two independent materials constants for homogeneous and isotropic materials.

We can also express stresses in terms of strains by solving the above equation,

The above relations are valid for plane stress case. For plane strain case, we need to replace the material constants in the above equations in the following fashion,

$$n \qquad \frac{n}{1 - n}$$

For example, the stress is related to strain by

Initial strains due to *temperature change* (thermal loading) is given by,

where a is the coefficient of thermal expansion, T the change of temperature. Note that if the structure is free to deform under thermal loading, there will be no (elastic) stresses in the structure.

3.6 GENERALIZED COORDINATES APPROACH TO NODEL APPROXIMATIONS

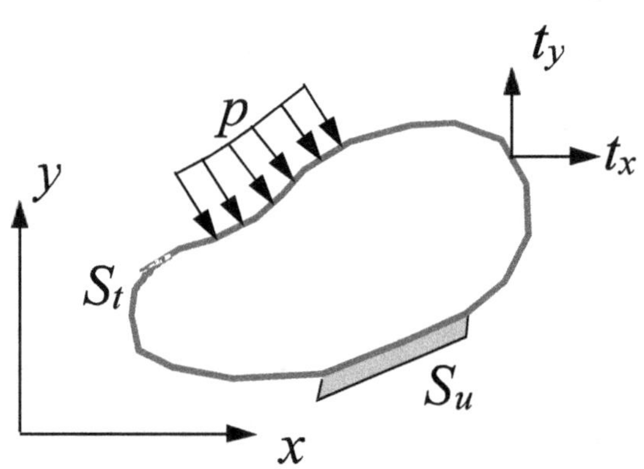

The boundary S of the body can be divided into two parts, S_u and S_t. The boundary conditions (BC's) are described as, in which t_x and t_y are traction forces (stresses on the boundary) and the barred quantities are those with known values.

In FEM, all types of loads (distributed surface loads, body forces, concentrated forces and moments, etc.) are converted to point forces acting at the nodes.

Exact Elasticity Solution

The exact solution (displacements, strains and stresses) of a given problem must satisfy the equilibrium equations, the given boundary conditions and compatibility conditions (structures should deform in a continuous manner, no cracks and overlaps in the obtained displacement field)

3.7 ISOPARAMETRIC ELEMENTS

In one dimensional problem, each node is allowed to move only in $\pm$ x direction. But in two dimensional problem, each node is permitted to move in the two directions i.e., x and y.

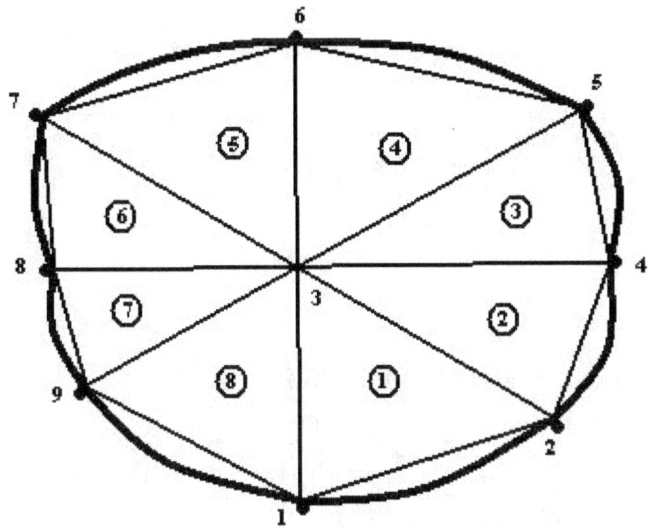

The element connectivity table for the above domain is explained as table.

Element (e)	Nodes
(1)	123
(2)	234
(3)	435
(4)	536
(5)	637
(6)	738
(7)	839
(8)	931

- **Constant Strain Triangular (CST) Element**

  A three noded triangular element is known as constant strain triangular (CST) element. It has six unknown displacement degrees of freedom (u_1v_1, u_2v_2, u_3v_3).

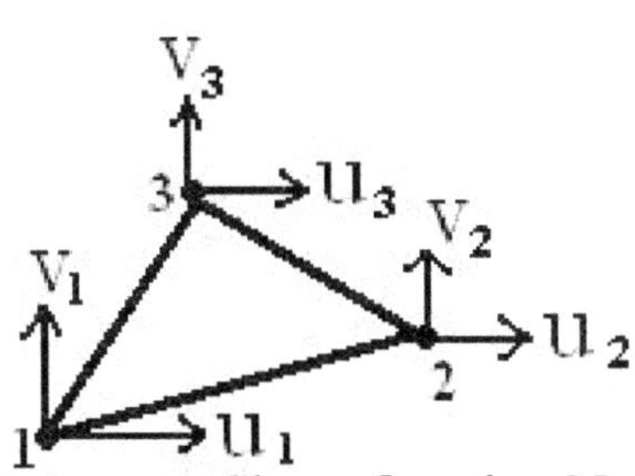

- **Shape function for the CST element** Shape function $N_1 = (p_1 + q_1x + r_1y) / 2A$

Shape function $N_2 = (p_2 + q_2x + r_2y) / 2A$

Shape function $N_3 = (p_3 + q_3x + r_3y) / 2A$

- **Displacement function for the CST element**

Displacement function u =

$$\begin{Bmatrix} u(x,y) \\ v(x,y) \end{Bmatrix} = \begin{bmatrix} N1 & 0 & N2 & 0 & N3 & 0 \\ 0 & N1 & 0 & N2 & 0 & N3 \end{bmatrix} X \begin{Bmatrix} u1 \\ v1 \\ u2 \\ v2 \\ u3 \\ v3 \end{Bmatrix}$$

- **Strain – Displacement matrix [B] for CST element**

Strain – Displacement matrix [B] =

$$\frac{1}{2A}\begin{bmatrix} q_1 & 0 & q_2 & 0 & q_3 & 0 \\ 0 & r_1 & 0 & r_2 & 0 & r_3 \\ r_1 & q_1 & r_2 & q_2 & r_3 & q_3 \end{bmatrix}$$

Where, $q_1 = y_2 - y_3$ $\quad r_1 = x_3 - x_2$

$q_2 = y_3 - y_1$ $\quad r_2 = x_1 - x_3$

$q_3 = y_1 - y_2$ $\quad r_3 = x_2 - x_1$

- **Stress – Strain relationship matrix (or) Constitutive matrix [D] for two dimensional element**

$$E \begin{bmatrix} 1-v & v & v & 0 & 0 & 0 \\ v & 1-v & v & 0 & 0 & 0 \\ v & v & 1-v & 0 & 0 & 0 \end{bmatrix}$$

- **Stress – Strain relationship matrix for two dimensional plane stress problems**

The normal stress σ_z and shear stresses τ_{xz}, τ_{yz} are zero.

$$[D] = \frac{E}{1-v^2}\begin{bmatrix} 1 & v & 0 \\ v & 1 & 0 \\ 0 & 0 & \frac{1-v}{2} \end{bmatrix}$$

- **Stress – Strain relationship matrix for two dimensional plane strain problems**

Normal strain e_z and shear strains e_{xz}, e_{yz} are zero.

- **Stiffness matrix equation for two dimensional element (CST element)**

Stiffness matrix $[k] = [B]^T [D] [B] A t$

$$[B] = \frac{1}{2A}\begin{bmatrix} q_1 & 0 & q_2 & 0 & q_3 & 0 \\ 0 & r_1 & 0 & r_2 & 0 & r_3 \\ r_1 & q_1 & r_2 & q_2 & r_3 & q_3 \end{bmatrix}$$

For plane stress problems,

$$[D] = \frac{E}{1-v^2}\begin{bmatrix} 1 & v & 0 \\ v & 1 & 0 \\ 0 & 0 & \frac{1-v}{2} \end{bmatrix}$$

For plane strain problems,

- **Temperature Effects**

Distribution of the change in temperature (ΔTi)s known as strain. Due to the change in temperature can be considered as an initial strain e_0.

$$\sigma = D\,(Bu - e_0)$$

- **Galerkin Approach**

Stiffness matrix $[K]_e = [B]^T [D][B] A t$.

Force Vector $\{F\}_e = [K]_e \{u\}$

- **Linear Strain Triangular (LST) element**

A six noded triangular element is known as Linear Strain Triangular (LST) element. It has twelve unknown displacement degrees of freedom. The displacement functions of the element are quadratic instead of linear as in the CST.

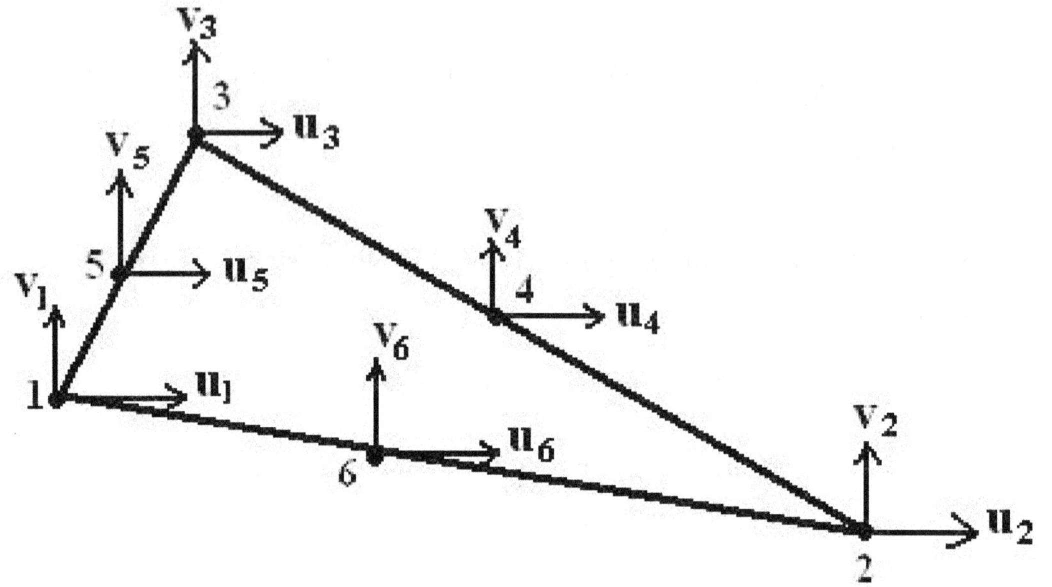

- **Problem (I set)**

1. Determine the shape functions N_1, N_2 and N_3 at the interior point P for the triangular element for the given figure.

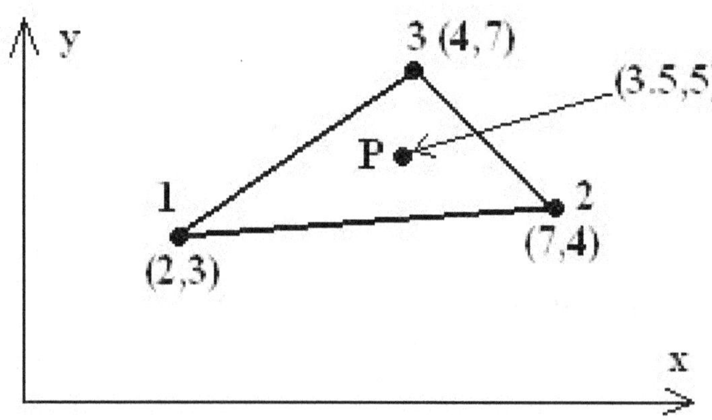

The two dimensional propped beam shown in figure. It is divided into two CST elements. Determine the nodal displacement and element stresses using plane stress conditions. Body force is neglected in comparison with the external forces.

Take, Thickness (t) = 10mm,

Young's modulus (E)= $2x10^5$ N/mm^2,

Poisson's ratio (v) = 0.25.

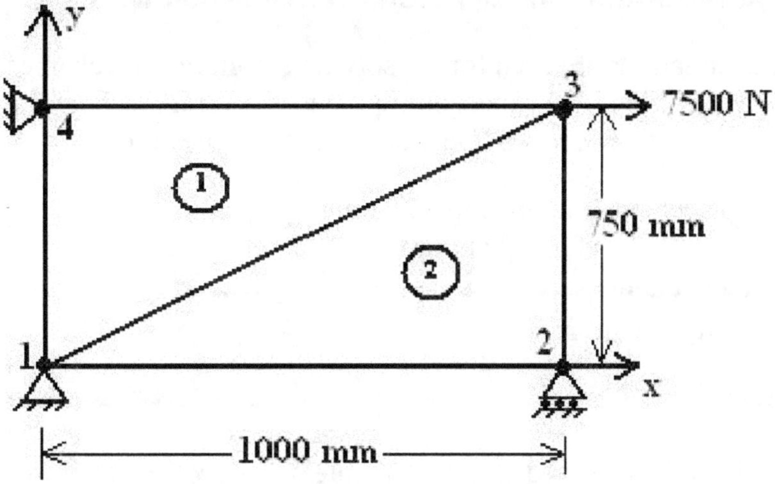

3. A thin plate is subjected to surface traction as in figure. Calculate the global stiffness matrix.

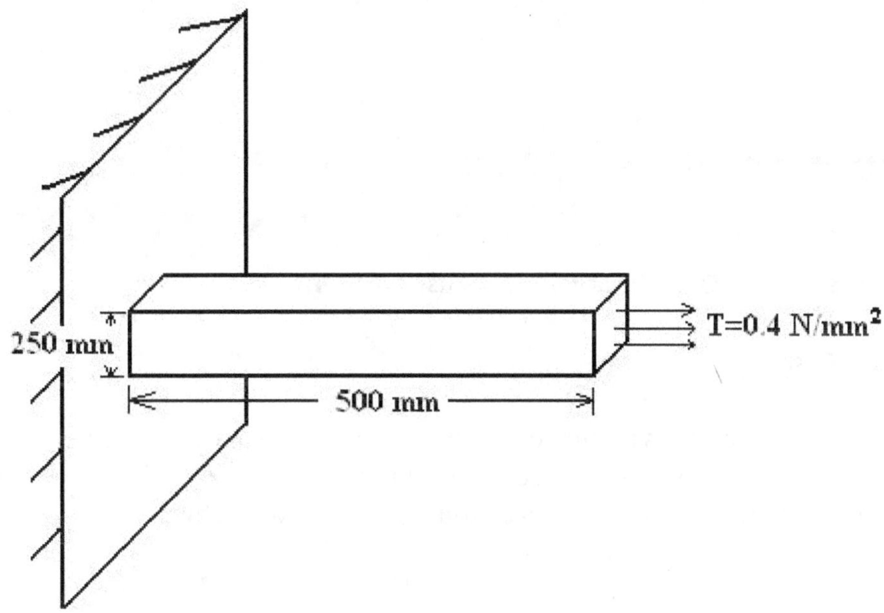

- **Scalar variable problems**

In structural problems, displacement at each nodal point is obtained. By using these displacement solutions, stresses and strains are calculated for each element. In structural problems, the unknowns (displacements) are represented by the components of vector field. For example, in a two dimensional plate, the unknown quantity is the vector field u(x, y), where u is a (2x1) displacement vector.

3.8 STRUCTURAL MECHANICS APPLICATIONS IN 2 DIMENSIONS

Elasticity equations are used for solving structural mechanics problems. These equations must be satisfied if an exact solution to a structural mechanics problem is to be obtained. Thest are four basic sets of elasticity equations they are

- Strain displacement relationship equations
- Stress strain relationship equations
- Equilibrium equations
- Compatibility equations

TRUSS ELEMENT

A truss element is defined as a deformable, two-force member that is subjected to loads in the axial direction. The loads can be tensile or compressive. The only degree of freedom for a one-dimensional truss (bar) element is axial (horizontal) displacement at each node.

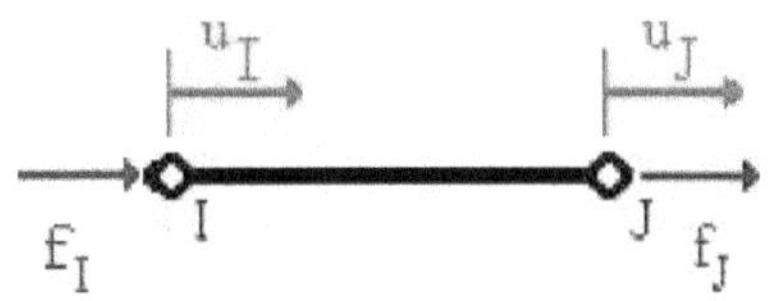

Assumptions for the One-Dimensional Truss Element

Prismatic Member

The truss element is assumed to have a constant cross-section, i.e., it is a prismatic member. If a truss structure is stepped, then it must be divided up into sections of constant cross-section in order to obtain an exact solution as shown below.

If a truss structure is tapered, then it can be approximated by using many small truss elements, each having the same cross-section as the middle of the tapered length it is approximating. The more sections that are used to approximate a tapered truss, the more accurate the solution will be.

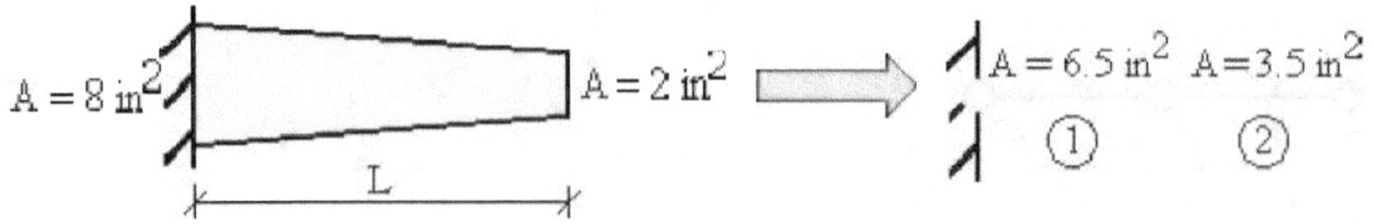

Weightless Member

The weight (W) of the truss is neglected since it is assumed to be much less than the total resultant forces (F) acting on the truss. If the weight of the truss is not neglected, then its effects must be represented as vertical forces acting at the nodes. But since truss element is defined as two-force member it cannot have any vertical (shear) force, thus the member weight has to be neglected. If shear forces exist, then a beam element must be used to model the structure.

Nodal Forces

For one-dimensional truss element, forces (loads) can only be applied at the nodes of the element, but not between the nodes. This is consistent with the FEM equations which relate nodal forces to nodal displacements through the stiffness matrix.

Axially Loaded

For one-dimensional truss element, forces (loads) can only be applied at the centroid of the element cross-sectional area.

Buckling Effect not Considered

A bar element can be subjected to either tensile or compressive forces. Tensile forces can be applied to a bar of any cross-sectional area or member length, and failure is associated with sudden fracture or general yielding. When compressive forces are applied to a member, it can either fail due to crushing or buckling. Buckling is present when the member bends and laterally deflects as shown on the right figure below.

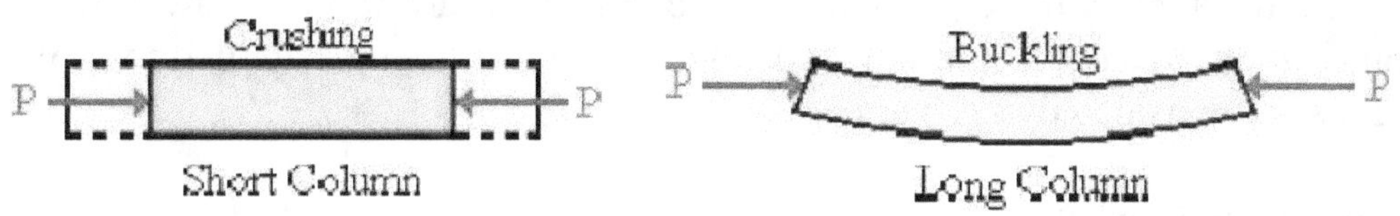

Buckling is not accounted for in the formulation of the truss element. Members that do not buckle are classified as short columns and members that buckles are classified as long columns. The structural response of a short column can be predicted with a truss element.

To determine if buckling will occur the reader should refer to a mechanics of material textbook. We will now introduce a simple geometric guideline to determine if buckling might occur. If the ratio between the member length and the least dimension of the cross-section is equal or less than 10, the member is considered a short column and buckling will not occur, i.e.,

$$\frac{L}{d} \leq 10$$

d - Least Cross-Sectional Dimension

L - Member Length

Two examples include

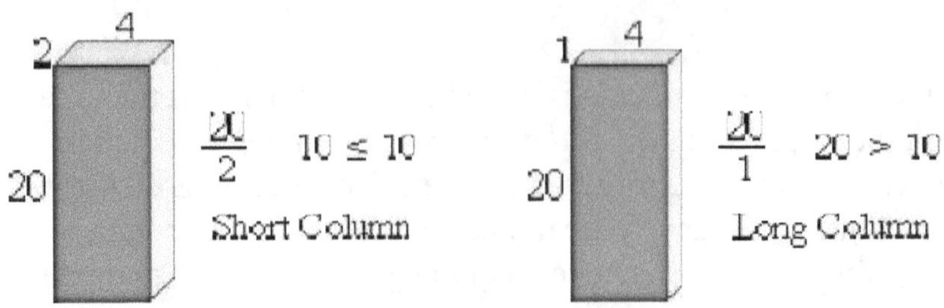

In the second case if a bar element is subjected to a compressive force, the element will not predict the buckling response. One should note that the above geometric rule is a simple guideline, however, in reality buckling depends not only on the member length and cross-sectional area, but material properties and support conditions.

Isotropic Material

A truss element has the same mechanical and physical properties in all directions, i.e., they are independent of direction. For instance, cutting out three tensile test specimens, one in the *x*-direction, one in the *y*-direction and the other oriented 45 degrees in the *x*-*y* plane, a tension test on each specimen, will result in the same mechanical values for the modulus of elasticity (E), yield strength τ_y and ultimate strength σ_u. Most metals are considered isotropic. In contrast fibrous materials, such as wood, typically have properties that are directionally dependant and are generally considered anisotropic (not isotropic).

Constant (Static) Load

The loads that are applied to the truss element are assumed to be static and not to vary over the time period being considered. This assumption is only valid if the rate of change of the force is much less than the applied force ($F >> dF/dt$), i.e., the loads are applied slowly. If the loads vary significantly, (if the variation in load is not much less than the applied force) then the problem must be considered as dynamic.

Poisson's Effect not Considered

Poisson's ratio is a material parameter. Poisson's effect is when a uniform cross-section bar is subject to a tensile load, and the axial stretching is accompanied by a contraction in the lateral dimension. For one-dimensional truss element., this effect is neglected for simplicity, i.e., $v = 0$.

Cross Section Remains Plane

For one-dimensional element, although the force(s) are acting on only the centroid of the truss (bar) element, it is assumed that it has a uniform effect to the plane. Thus the cross section will move uniformly and remain plane and normal to the axial axis before and after loading.

Homogenous Material

A truss element has the same material composition throughout and therefore the same mechanical properties at every position in the material. Therefore, the modulus of elasticity E is constant throughout the truss element. A member in which the material properties varies from one point to the next in the member is called inhomogenous (non-homogenous). If a truss is composed of different types of materials, then it must be divide up into elements that are each of a single homogeneous material, otherwise the solution will not be exact.

The left figure shows a composite bar composed of brass and aluminum. This structure can be divided into two elements as shown on the right, one element for the brass with $E_1 = 15 \times 10^6$ psi and one for the aluminum with $E_2 = 10 \times 10^6$ psi.

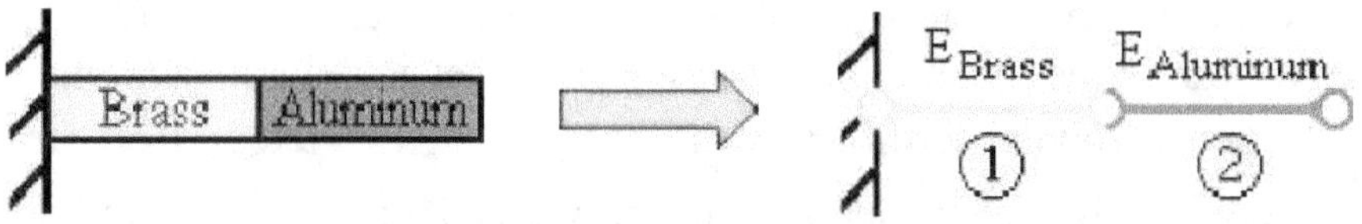

TRUSS ELEMENT (OR SPAR ELEMENT OR LINK ELEMENT)

Differentiate between a truss and a frame.

Truss	Frame
Only concentrated loads act.	Concentrated loads, uniformly distributed loads, moments, all can act.
Loads act only at the joints.	Loads can be applied at the joints and/or in-between the joints
Truss members undergo only axial deformation (along the length of the member).	Frame members can undergo axial and bending deformations (translations as well as rotations).

A **grid** is a structure on which loads applied perpendicular to the plane of the structure, as opposed to a plane frame, where loads are applied in the plane of the structure.

6.7.1 Derivation of stiffness matrix and finite element equation for a truss element.

There are two joints for an arbitrarily inclined single truss element (at an angle θ, positive counter-clockwise from +ve *x*-axis). For each joint *i*, there are two degrees of freedom, i.e., a joint can have horizontal displacement (u_i) and vertical displacement (v_i). Hence, for a single truss element, there are 4 degrees of freedom. The nodal displacement degrees of freedom and the nodal force degrees of freedom are shown in the following figure.

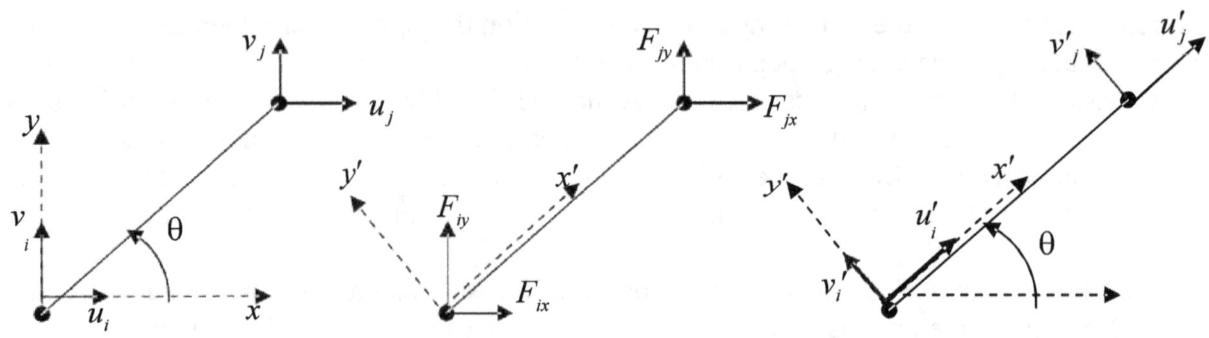

Note that the deformations occurring in the truss members are so small that they are only axial. The axial displacement of the truss can be resolved along horizontal x-axis and vertical y-axis. But in our derivation, let us resolve the horizontal and vertical displacements (in xy-axes) of a joint along and perpendicular to the truss member (in $x'y'$-axes). Refer to the Figure in the next page. Note $u_i \sin\theta$ component acting towards negative y'-direction and all other components acting towards in +ve x'- and y'-directions.

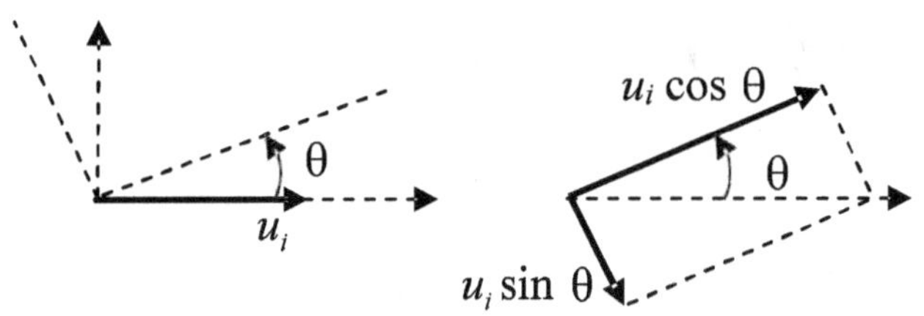

$$u'_i = u_i \cos\theta + v_i \sin\theta$$
$$v'_i = -u_i \sin\theta + v_i \cos\theta$$

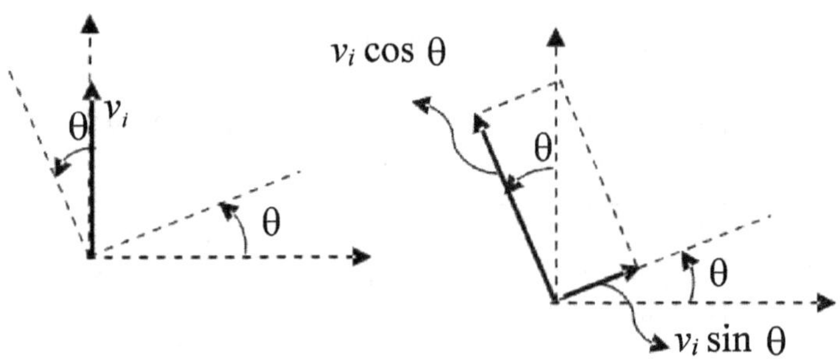

$$u'_j = u_j \cos\theta + v_j \sin\theta$$
$$v'_j = -u_j \sin\theta + v_j \cos\theta$$

The above equations can be written in the matrix form as follows

$$\begin{Bmatrix} u'_i \\ v'_i \\ u'_j \\ v'_j \end{Bmatrix} = \begin{bmatrix} \cos\theta & \sin\theta & 0 & 0 \\ -\sin\theta & \cos\theta & 0 & 0 \\ 0 & 0 & \cos\theta & \sin\theta \\ 0 & 0 & -\sin\theta & \cos\theta \end{bmatrix} \begin{Bmatrix} u_i \\ v_i \\ u_j \\ v_j \end{Bmatrix}$$

$\{u'\} = [T]\,\{u\}$ where $[T]$ is the transformation matrix

It is important to note that the displacements v'_i and v'_j are both zero since there can be no displacements perpendicular to the length of the member. Also $[T]^{-1} = [T]^T$

Similarly, we resolve forces along the length of the member (positive x' direction) and perpendicular to the length of the member (positive y' direction)

$$\begin{Bmatrix} F'_{ix} \\ F'_{iy} \\ F'_{jx} \\ F'_{jy} \end{Bmatrix} = \begin{bmatrix} \cos\theta & \sin\theta & 0 & 0 \\ -\sin\theta & \cos\theta & 0 & 0 \\ 0 & 0 & \cos\theta & \sin\theta \\ 0 & 0 & -\sin\theta & \cos\theta \end{bmatrix} \begin{Bmatrix} F_{ix} \\ F_{iy} \\ F_{jx} \\ F_{jy} \end{Bmatrix}$$

$\{F\}' = [T]\{F\}$ where $[T]$ is the transformation matrix

The arbitrarily inclined truss member can be thought of as a simple bar element oriented at the same angle θ. Hence, we can write the finite element equation for this inclined bar element (in $x'y'$ coordinate system) as

$$\begin{Bmatrix} F'_{ix} \\ F'_{iy} \\ F'_{jx} \\ F'_{jy} \end{Bmatrix} = \frac{AE}{L}\begin{bmatrix} 1 & 0 & -1 & 0 \\ 0 & 0 & 0 & 0 \\ -1 & 0 & 1 & 0 \\ 0 & 0 & 0 & 0 \end{bmatrix} \begin{Bmatrix} u'_i \\ v'_i \\ u'_j \\ v'_j \end{Bmatrix}$$

$$\{F\}' = [k']\{u'\}$$

Substituting $\{F'\}$ and $\{u'\}$ from the previous equations, we can write

$$[T]\{F\} = [k'][T]\{u\}$$

Pre-multiplying the above equation by $[T]^{-1}$,

$$[T]^{-1}[T]\{F\} = [T]^{-1}[k'][T]\{u\}$$

But $[T]^{-1}[T] = 1$ and the above equation can be written as

$$\{F\} = [k]\{u\} \quad \text{where } [k] = [T]^{-1}[k'][T]$$

Carrying out the matrix multiplication for $[k]$, we obtain

$$\begin{Bmatrix} F_{ix} \\ F_{iy} \\ F_{jx} \\ F_{jy} \end{Bmatrix} = \frac{AE}{L}\begin{bmatrix} c^2 & cs & -c^2 & -cs \\ cs & s^2 & -cs & -s^2 \\ -c^2 & -cs & c^2 & cs \\ -cs & -s^2 & cs & s^2 \end{bmatrix} \begin{Bmatrix} u_i \\ v_i \\ u_j \\ v_j \end{Bmatrix}$$

where $c = \cos^2\theta$ and $s = \sin^2\theta$.

Computation of strain and stress in the truss element

The change in length of the truss member is equal to the change in axial displacement of the truss member in the $x'y'$ co-ordinate system

$$\delta = u'_j - u'_i$$

$$\delta = (u_j \cos\theta + v_j \sin\theta) - (u_i \cos\theta + v_i \sin\theta)$$

$$\delta = \langle \cos\theta \quad -\sin\theta \quad \cos\theta \quad \sin\theta \rangle \begin{Bmatrix} u_i \\ v_i \\ u_j \\ v_j \end{Bmatrix}$$

Strain in the truss element is given by $\varepsilon^e = \frac{\delta}{L}$, i.e.,

$$\varepsilon^e = \frac{\langle -\cos\theta - \sin\theta \quad \cos\theta \quad \sin\theta \rangle}{L} \begin{Bmatrix} u_i \\ v_i \\ u_j \\ v_j \end{Bmatrix}$$

Stress in the truss element is given by $\sigma^e = E\,\varepsilon^e$, i.e.,

$$\sigma^e = E\frac{\langle -\cos\theta - \sin\theta \quad \cos\theta \quad \sin\theta \rangle}{L} \begin{Bmatrix} u_i \\ v_i \\ u_j \\ v_j \end{Bmatrix}$$

Problem

The two-element truss is subjected to external loading as shown in figure. Using the same node and element numbering as shown in figure, determine the displacement components at node 3, the reaction components at nodes 1 and 2, and the element displacement, stresses and forces. The elements have modulus of elasticity $E_1 = E_2 = 10\times10^6$ $\frac{lb}{in^2}$ and cross-sectional areas $A_1 = A_2 = 1.5\ in^2$

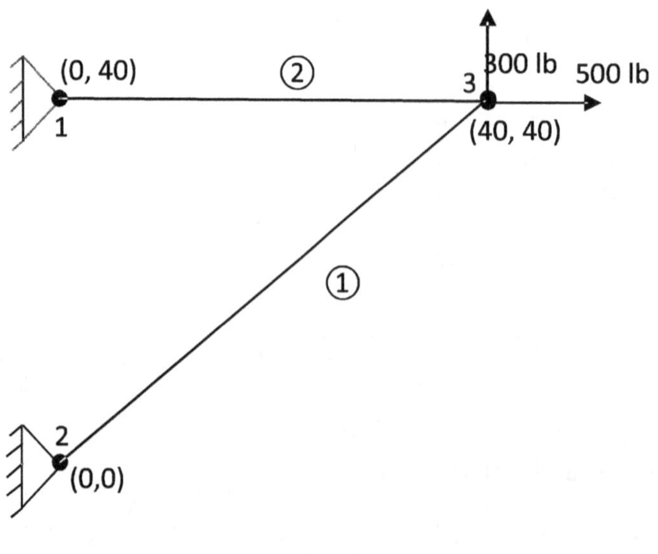

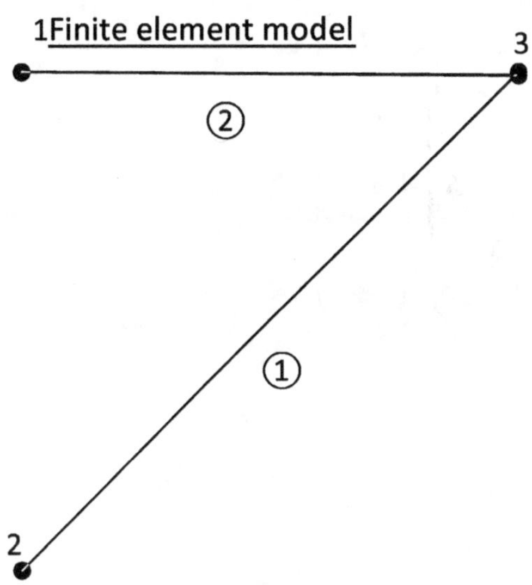

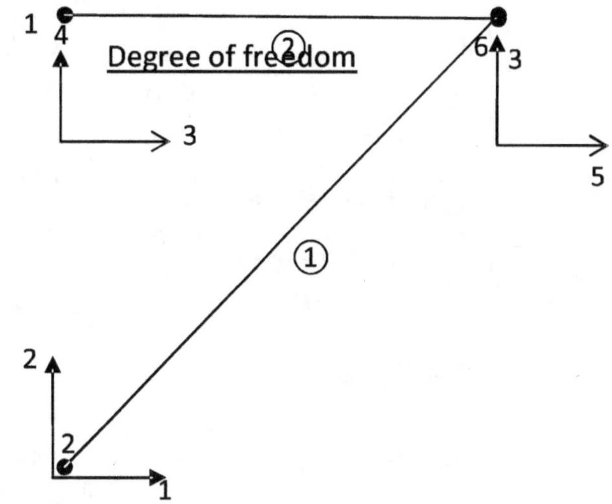

For element 1

For element 2

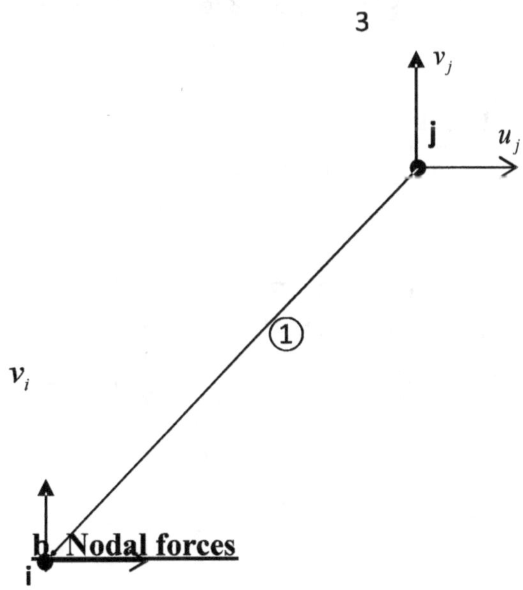

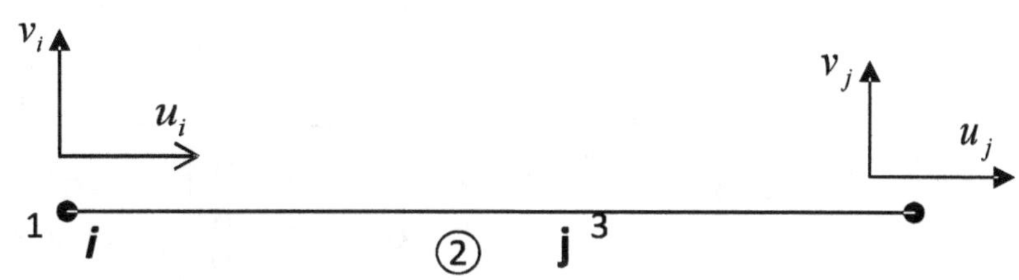

b. Nodal forces

element 1

element 2

FINITE ELEMENT EQUATION

$$\begin{pmatrix} F_{ix} \\ F_{iy} \\ F_{jx} \\ F_{jy} \end{pmatrix} = \frac{AE}{L} \begin{pmatrix} c^2 & c\,s & -c^2 & -c\,s \\ c\,s & s^2 & -c\,s & -s^2 \\ -c^2 & -c\,s & c^2 & c\,s \\ -c\,s & s^2 & c\,s & s^2 \end{pmatrix} \begin{pmatrix} u_i \\ v_i \\ u_j \\ v_j \end{pmatrix}$$

For element 1

$$\theta = 45, \qquad \frac{AE}{L} = \frac{1.5 \times 10 \times 10^6}{56.5685} = 2.6516 \times 10^5 \frac{lb}{in^2}$$

$$[K]^{(1)} = \frac{AE}{L} \begin{pmatrix} 0.5 & 0.5 & -0.5 & -0.5 \\ 0.5 & 0.5 & -0.5 & -0.5 \\ -0.5 & -0.5 & 0.5 & 0.5 \\ 0.5 & -0.5 & 0.5 & 0.5 \end{pmatrix} = 10^5 \begin{array}{c} 1 \\ 2 \\ 5 \\ 6 \end{array} \overset{\begin{array}{cccc} 1 & 2 & 5 & 6 \end{array}}{\begin{pmatrix} 1.325826 & 1.325826 & -1.325826 & -1.325826 \\ 1.325826 & 1.325826 & -1.325826 & -1.325826 \\ -1.325826 & -1.325826 & 1.325826 & 1.325826 \\ -1.325826 & -1.325826 & 1.325826 & 1.325826 \end{pmatrix}}$$

$$\theta = 0, \qquad \frac{AE}{L} = \frac{1.5 \times 10 \times 10^6}{40} = 3.75 \times 10^5$$

For element 2

$$[K]^{(2)} = \frac{AE}{L} \begin{bmatrix} 1 & 0 & -1 & 0 \\ 0 & 0 & 0 & 0 \\ -1 & 0 & 1 & 0 \\ 0 & 0 & 0 & 0 \end{bmatrix} = 10^5 \begin{array}{c} 3 \\ 4 \\ 5 \\ 6 \end{array} \overset{\begin{array}{cccc} 3 & 4 & 5 & 6 \end{array}}{\begin{bmatrix} 3.75 & 0 & -3.75 & 0 \\ 0 & 0 & 0 & 0 \\ -3.75 & 0 & 3.75 & 0 \\ 0 & 0 & 0 & 0 \end{bmatrix}}$$

Assembly of finite element equation

$$\begin{pmatrix} F_{1x} \\ F_{1y} \\ F_{2x} \\ F_{2y} \\ F_{3x} \\ F_{3y} \end{pmatrix} = 10^5 \begin{array}{c} 1 \\ 2 \\ 3 \\ 4 \\ 5 \\ 6 \end{array} \overset{\begin{array}{cccccc} 1 & 2 & 3 & 4 & 5 & 6 \end{array}}{\begin{pmatrix} 1.325 & 1.325 & 0 & 0 & -1.325 & -1.325 \\ 1.325 & 1.325 & 0 & 0 & -1.325 & -1.325 \\ 0 & 0 & 3.75 & 0 & -3.75 & 0 \\ 0 & 0 & 0 & 0 & 0 & 0 \\ -1.325 & -1.325 & -3.75 & 0 & 5.0751 & 1.325 \\ -1.325 & -1.325 & 0 & 0 & 1.325 & 1.325 \end{pmatrix}} \begin{pmatrix} u_1 \\ v_1 \\ u_2 \\ v_2 \\ u_3 \\ v_3 \end{pmatrix}$$

Applying boundary conditions

$F_{3x} = 500$ lb

$F_{3y} = 300$ lb

$u_1 = 0$

$u_2 = 0$

$v_1 = 0$

$v_2 = 0$

knowns	unknowns
$F_{3x} = 500$ lb	F_{1x}
$F_{3y} = 300$ lb	F_{1y}
$u_1 = 0$	F_{2x}
$v_1 = 0$	F_{2y}
$u_2 = 0$	u_3
$v_2 = 0$	v_3

$$\begin{Bmatrix} F_{1x} \\ F_{1y} \\ F_{2x} \\ F_{2y} \\ 500 \\ 300 \end{Bmatrix} = 10^5 \begin{bmatrix} 1.325826 & 1.325826 & 0 & 0 & -1.325826 & -1.325826 \\ 1.325826 & 1.325826 & 0 & 0 & -1.325826 & -1.325826 \\ 0 & 0 & 3.75 & 0 & -3.75 & 0 \\ 0 & 0 & 0 & 0 & 0 & 0 \\ -1.325826 & -1.325826 & -3.75 & 0 & 5.0751826 & 1.325826 \\ -1.325826 & -1.325826 & 0 & 0 & 1.325826 & 1.325826 \end{bmatrix} \begin{Bmatrix} 0 \\ 0 \\ 0 \\ 0 \\ u_3 \\ v_3 \end{Bmatrix}$$

Eliminating the 1st, 2nd, 3rd and 4th column to solve u_3 and v_3

$$500 = 10^5 (5.075u_3 + 1.325v_3) \longrightarrow (1)$$

$$300 = 10^5 (1.325u_3 + 1.325v_3) \longrightarrow (2)$$

Solve the equation (1) and (2)

$$\boxed{u_3 = 5.33 \times 10^{-4} \ in}$$

$$\boxed{v_3 = 1.731 \times 10^{-3} in}$$

$F_{1x} = -300\ lb$

$F_{1y} = -300\ lb$

$F_{2x} = -200\ lb$

$F_{2y} = 0$

Finding element stresses

For element 1

$$\sigma_1 = \frac{E}{L}[u_5 \cos 0^\circ - u_3 \sin 0^\circ] = \frac{10\times10^6}{40}[(0.5333\times1) - (0\times0)] = 133.325 \frac{lb}{in^2}$$

$$\sigma_2 = \frac{E}{L}\left[(u_5 \cos 45^\circ - u_6 \sin 45^\circ) - (u_1 \cos 45^\circ + u_2 \sin 45^\circ)\right]$$

$$= \frac{10\times10^6}{56.57}\left[(0.5333\times10^{-3} \cos 45^\circ) + (1.731\times10^{-3} \sin 45^\circ) - 0\right]$$

$$= \frac{10\times10^6}{56.57}[0.0003771 + 0.001224] = 283.03 \frac{lb}{in^2}$$

PROBLEM

To illustrate how we can combine spring and bar element in one structure, we can solve the two-bar truss supported by a spring as shown below. Both bars have E = 210 GPa and A = 5.0 x10 $^{-4}$ m^2. Bar one has a length of 5 m and bar two a length of 10 m. the spring stiffness is k = 2000 kN/m.

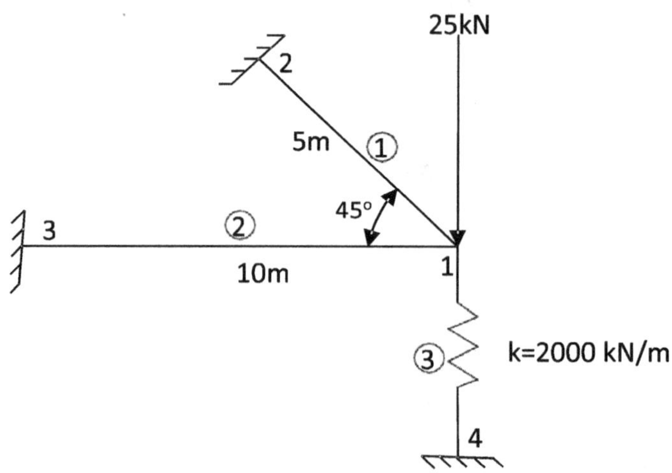

Solution :

Given : E = 210 GPa

$A = 5.0 \text{ x}10^{-4}$ m^2 , $L_1 = 5$ m, $L_2 = 10$ m, $K = 2 \text{ x } 10^6 \frac{N}{m^2}$

NOTE: A spring is considered as a bar element whose stiffness is $2 \text{ x } 10^6 \frac{N}{m^2}$

STEP 1 : Finite Element Representation Of Forces And Displacements

Displacements

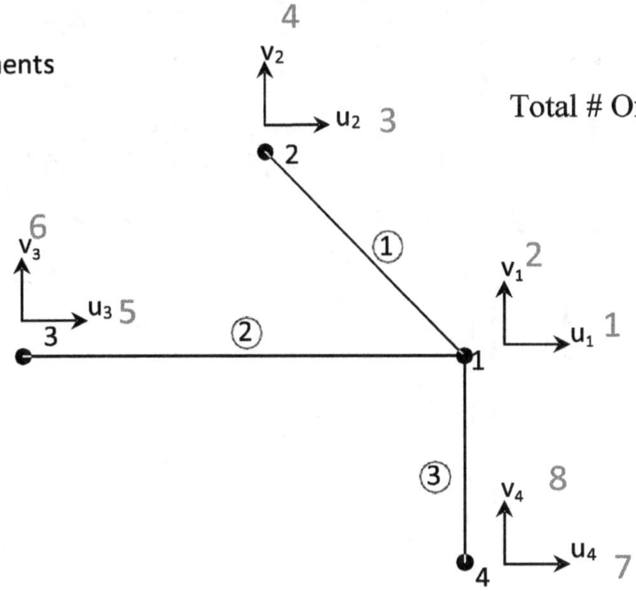

Total # Of Degrees Of Freedom: 8

Forces

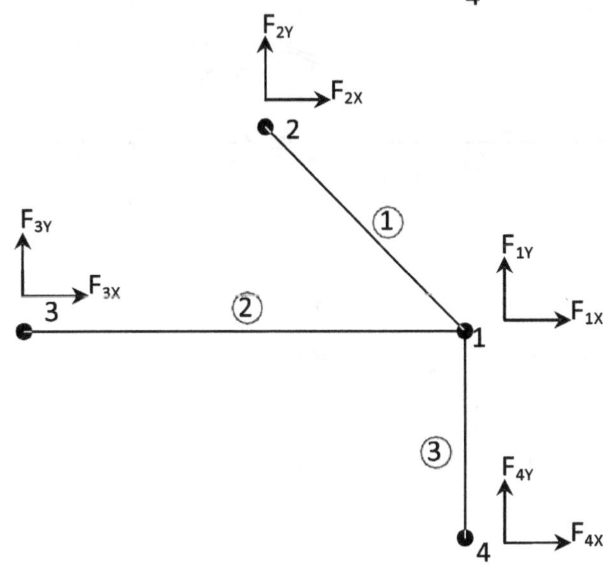

Step 2: Finite Element Equations

Element 1:

$\theta=135^0$

$l^2=\cos^2\theta=0.5$

$m^2=\sin^2\theta=0.5$

$lm=\cos\theta\sin\theta=-0.5$

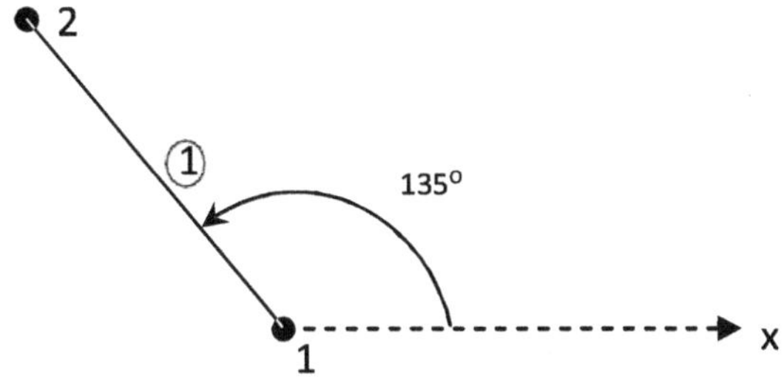

$$[K]^{(1)} = \frac{\left(5\times10^{-4}\,\text{m}^2\right)\left(210\times10^{6}\,\text{kN/m}^2\right)}{5\text{ m}} \begin{array}{c} \\ 1 \\ 2 \\ 3 \\ 4 \end{array} \begin{array}{cccc} 1 & 2 & 3 & 4 \\ \hline 0.5 & -0.5 & -0.5 & 0.5 \\ -0.5 & 0.5 & 0.5 & -0.5 \\ -0.5 & 0.5 & 0.5 & -0.5 \\ 0.5 & -0.5 & -0.5 & 0.5 \end{array}$$

$$[K]^{(1)} = 105\times10^{5} \begin{array}{c} \\ 1 \\ 2 \\ 3 \\ 4 \end{array} \begin{array}{cccc} 1 & 2 & 3 & 4 \\ \hline 1 & -1 & -1 & 1 \\ -1 & 1 & 1 & -1 \\ -1 & 1 & 1 & -1 \\ 1 & -1 & -1 & 1 \end{array}$$

Element 2:

$\theta = 180^0$

$l^2 = \cos^2\theta = 1$

$m^2 = \sin^2\theta = 0$

$lm = \cos\theta \sin\theta = 0$

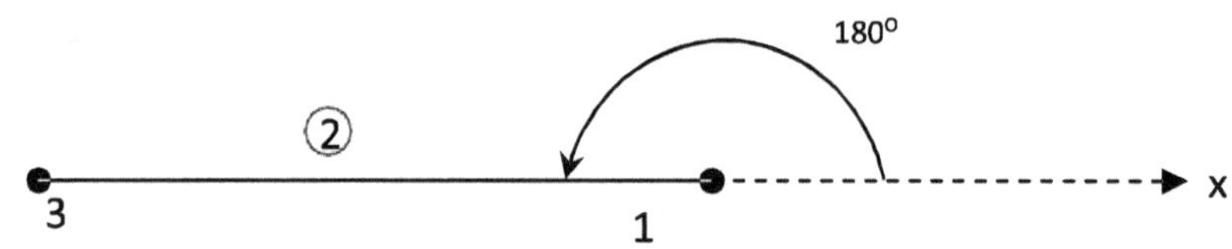

$$[K]^{(2)} = \frac{\left(5\times10^{-4}\,\text{m}^2\right)\left(210\times10^{6}\,\text{kN/m}^2\right)}{10\text{ m}} \begin{array}{c} \\ 1 \\ 2 \\ 3 \\ 4 \end{array} \begin{array}{cccc} 1 & 2 & 5 & 6 \\ \hline 1 & 0 & -1 & 0 \\ 0 & 0 & 0 & 0 \\ -1 & 0 & 1 & 0 \\ 0 & 0 & 0 & 0 \end{array}$$

$$[K]^{(2)} = 105\times10^{5} \begin{array}{c} \\ 1 \\ 2 \\ 5 \\ 6 \end{array} \begin{array}{cccc} 1 & 2 & 5 & 6 \\ \hline 1 & 0 & -1 & 0 \\ 0 & 0 & 0 & 0 \\ -1 & 0 & 1 & 0 \\ 0 & 0 & 0 & 0 \end{array}$$

Element 3:

$\theta = 270^0$

$l^2 = \cos^2\theta = 0$

$m^2 = \sin^2\theta = 1$

$lm = \cos\theta \sin\theta = 0$

$$
[K]^{(3)} = 2 \times 10^{6}
\begin{array}{c}
\\ 1 \\ 2 \\ 7 \\ 8
\end{array}
\!\!
\begin{array}{c}
\begin{array}{cccc} 1 & 2 & 7 & 8 \end{array} \\
\begin{bmatrix}
0 & 0 & 0 & 0 \\
0 & 1 & 0 & -1 \\
0 & 0 & 0 & 0 \\
0 & -1 & 0 & 1
\end{bmatrix}
\end{array}
$$

STEP 3: Combination Of Finite Element Equations

$$
\begin{Bmatrix} F_{1X} \\ F_{1Y} \\ F_{2X} \\ F_{2Y} \\ F_{3X} \\ F_{3Y} \\ F_{4X} \\ F_{4Y} \end{Bmatrix}
= 10^{5} \times
\begin{array}{c}
\\ 1 \\ 2 \\ 3 \\ 4 \\ 5 \\ 6 \\ 7 \\ 8
\end{array}
\!\!
\begin{array}{c}
\begin{array}{cccccccc} 1 & 2 & 3 & 4 & 5 & 6 & 7 & 8 \end{array} \\
\begin{bmatrix}
210 & -105 & -105 & 105 & -105 & 0 & 0 & 0 \\
-105 & 125 & 105 & -105 & 0 & 0 & 0 & -20 \\
-105 & 105 & 105 & -105 & 0 & 0 & 0 & 0 \\
105 & -108 & -105 & 105 & 0 & 0 & 0 & 0 \\
-105 & 0 & 0 & 0 & 105 & 0 & 0 & 0 \\
0 & 0 & 0 & 0 & 0 & 0 & 0 & 0 \\
0 & 0 & 0 & 0 & 0 & 0 & 0 & 0 \\
0 & -20 & 0 & 0 & 0 & 0 & 0 & 20
\end{bmatrix}
\end{array}
\begin{Bmatrix} u_1 \\ v_1 \\ u_2 \\ v_2 \\ u_3 \\ v_3 \\ u_4 \\ v_4 \end{Bmatrix}
$$

STEP 4: Applying Boundary Conditions:
Since nodes 1, 2, and 3 are fixed, we have

$u_2 = v_2 = 0$; $u_3 = v_3 = 0$; $u_4 = v_4 = 0$;

$F_{1x} = 0$ and $F_{1y} = -25$ kN

$$
\begin{Bmatrix} 0 \\ -25 \\ F_{2X} \\ F_{2Y} \\ F_{3X} \\ F_{3Y} \\ F_{4X} \\ F_{4Y} \end{Bmatrix}
= 10^{5} \times
\begin{bmatrix}
210 & -105 & -105 & 105 & -105 & 0 & 0 & 0 \\
-105 & 125 & 105 & -105 & 0 & 0 & 0 & -20 \\
-105 & 105 & 105 & -105 & 0 & 0 & 0 & 0 \\
105 & -108 & -105 & 105 & 0 & 0 & 0 & 0 \\
-105 & 0 & 0 & 0 & 105 & 0 & 0 & 0 \\
0 & 0 & 0 & 0 & 0 & 0 & 0 & 0 \\
0 & 0 & 0 & 0 & 0 & 0 & 0 & 0 \\
0 & -20 & 0 & 0 & 0 & 0 & 0 & 20
\end{bmatrix}
\begin{Bmatrix} u_1 \\ v_1 \\ 0 \\ 0 \\ 0 \\ 0 \\ 0 \\ 0 \end{Bmatrix}
$$

Check whether there are as many unknowns as knowns.

STEP 5: SOLVING THE EQUATIONS:

Reduced matrix:

$$\begin{Bmatrix} 0 \\ -25 \end{Bmatrix} = 10^5 \times \begin{bmatrix} 210 & -105 \\ -105 & 125 \end{bmatrix} \begin{Bmatrix} u_1 \\ v_1 \end{Bmatrix}$$

On solving,

u_1=-1.724 x 10^{-3} m v_1=-3.4482 x 10^{-3} m

Find the reactions at supports by substituting the known nodal values

F_{2x} = -18.104 kN F_{2y} = 18.1041 kN

F_{3x} = 18.102 kN F_{3y} = 0

F_{4x} = 0 F_{4y} = 6.89 kN

STEP 6: Post Processing

Stress in element 1:

$$\sigma^{(1)} = \frac{E}{L}\begin{bmatrix} -l & -m & l & m \end{bmatrix} 10^{-3} \begin{Bmatrix} -1.724 \\ -3.4482 \\ 0 \\ 0 \end{Bmatrix}$$

$$\sigma^{(1)} = 51.2 \text{ MPa (Tensile)}$$

Stress in element 2:

$$\sigma^{(1)} = \frac{E}{L}\begin{bmatrix} -l & -m & l & m \end{bmatrix} 10^{-3} \begin{Bmatrix} -1.724 \\ -3.4482 \\ 0 \\ 0 \end{Bmatrix}$$

$$\sigma^{(2)} = -36.2 \text{ MPa (Compressive)}$$

PROBLEM

A circular concrete beam structure is loaded as shown. Find the deflection of points at 8”,16”, and the end of the beam. E = 4 x 106 psi

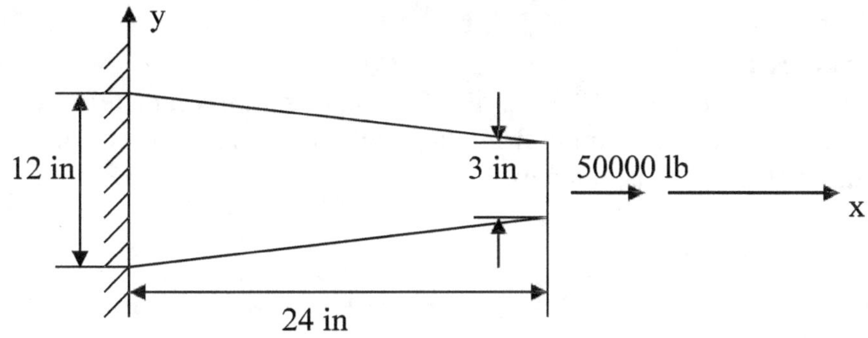

Solution

The beam structure looks very different from a spring. However, its behavior is very similar. Deflection occurs along the x-axis only. The only significant difference between the beam and a spring is that the beam has a variable cross-sectional area. An exact solution can be found if the beam is divided into an infinite number of elements, then, each element can be considered as a constant cross-section spring element, obeying the relation F = ku, where k is the stiffness constant of a beam element and is given by k = AE/L.

In order to keep size of the matrices small (for hand- calculations), let us divide the beam into only three elements. For engineering accuracy, the answer obtained will be in an acceptable range. If needed, accuracy can be improved by increasing the number of elements.

As mentioned earlier in this chapter, spring, truss, and beam elements are line-elements and the shape of the cross section of an element is irrelevant. Only the cross-sectional area is needed (also, moment of inertia for a beam element undergoing a bending load need to be defined). The beam elements and their computer models are shown

Here, the question of which cross-sectional area to be used for each beam section arises. A good approximation would be to take the diameter of the mid-section and use that to approximate the area of the element.

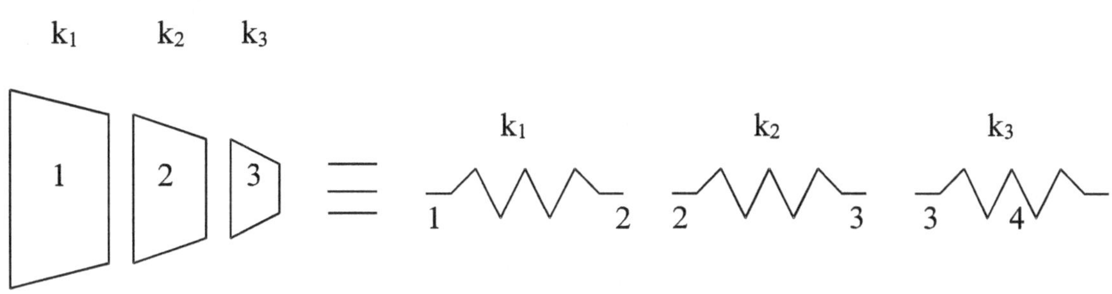

Cross-sectional area

The average diameters are: d_1 = 10.5 in., d_2 = 7.5 in., d_3 = 4.5. (diameters are taken at the mid sections and the values are found from the height and length ratio of the triangles shown in figure 2.10), which is given as

$12/L = 3/(L-24)$, $L = 32$

Average areas are:

A_1 = 86.59 in2 A_2 = 56.25 in2 A_3 = 15.9 in2

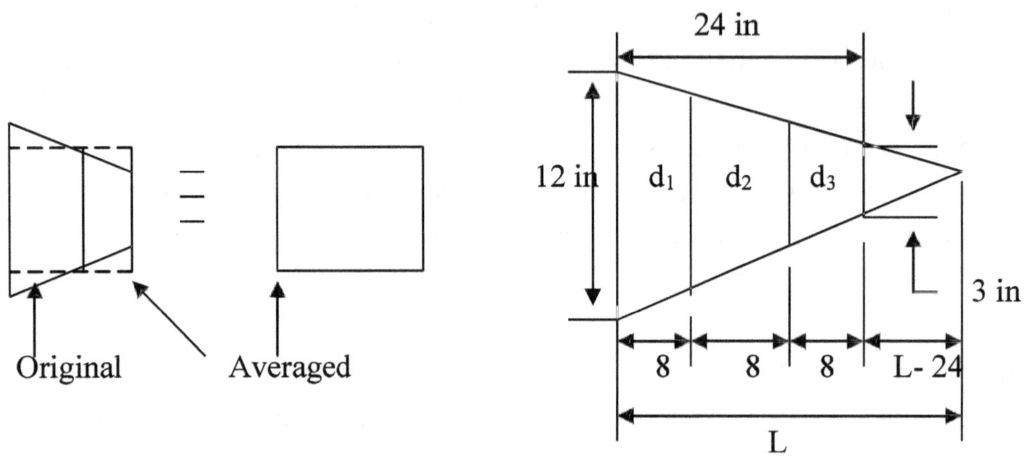

Stiffness

$k_1 = A_1 E/L_1 = (86.59)(4 \times 10^6/8) = 4.3295 \times 10^7$ lb./in., similarly,

$k_2 = A_2 E/L_2 = 2.8125 \times 10^7$ lb./in.

$k_3 = A_3 E/L_3 = 7.95 \times 10^6$ lb./in.

Element Stiffness Equations

$$[K^{(1)}] = 43.295 \times 10^7 \begin{pmatrix} 1 & -1 \\ -1 & 1 \end{pmatrix}$$

Similarly,

$$[K^{(2)}] = 28.125 \times 10^6 \begin{pmatrix} 1 & -1 \\ -1 & 1 \end{pmatrix}$$

$$[K^{(3)}] = 7.9500 \times 10^6 \begin{pmatrix} 1 & -1 \\ -1 & 1 \end{pmatrix}$$

Global stiffness matrix is

$$[K_g] = \begin{pmatrix} 43.295 & -43.295 & 0 & 0 \\ -43.295 & 43.295+28.125 & -28.125 & 0 \\ 0 & -28.125 & 28.125+7.95 & -7.95 \\ 0 & 0 & -7.95 & 7.95 \end{pmatrix} 10^6$$

Now the global structural equations can be written as,

$$10^6 \begin{pmatrix} 43.295 & -43.295 & 0 & 0 \\ -43.295 & 71.42 & -28.125 & 0 \\ 0 & -28.125 & 36.075 & -7.95 \\ 0 & 0 & -7.95 & 7.95 \end{pmatrix} \begin{Bmatrix} u_1 \\ u_2 \\ u_3 \\ u_4 \end{Bmatrix} = \begin{Bmatrix} F1 \\ F2 \\ F3 \\ F4 \end{Bmatrix}$$

Applying the boundary conditions: $u_1 = 0$, and $F_1 = F_2 = F_3 = 0$, $F_4 = 5000$ lb., results in the reduced matrix,

$$10^6 \begin{pmatrix} 71.42 & -28.125 & 0 \\ -28.125 & 36.075 & -7.95 \\ 0 & -7.95 & 7.95 \end{pmatrix} \begin{Bmatrix} u_2 \\ u_3 \\ u_4 \end{Bmatrix} = \begin{Bmatrix} 0 \\ 0 \\ 5000 \end{Bmatrix}$$

Solving we get,

$$\begin{Bmatrix} u_2 \\ u_3 \\ u_4 \end{Bmatrix} = \begin{Bmatrix} 0.0012 \\ 0.0029 \\ 0.0092 \end{Bmatrix} \text{ in.}$$

The deflections u_2, u_3, and u_4 are only the approximate values, which can be improved by dividing the beam into more elements. As the number of elements increases, the accuracy will improve.

UNIT IV

DYNAMIC ANALYSIS USING ELEMENT METHOD

4.1 INTRODUCTION

It provides the basic equations necessary for structural dynamical analysis and developed both the lumped and the consistent mass matrix involved in the analysis of bar beam and spring elements.

4.1.1 Fundamentals of Vibration

Any motion which repeats itself after an interval of time is called vibration or oscillation or periodic motion

All bodies possessing mass and elasticity are capable of producing vibration.

4.1.2 Causes of Vibrations

- Unbalanced forces in the machine. These force are produced from within the machine itself
- Elastic nature of the system.
- Self excitations produced by the dry friction between the two mating surfaces.
- External excitations applied on the system.
- Wind may causes vibrations
- Earthquakes may causes vibrations

4.1.3 Types of Vibrations

1. According to the actuating force

    Free or natural vibrations

    Forced vibrations

    Damped vibrations

    Undamped vibrations

2. According to motion of system with respect to axis

    Longitudinal vibrations

    Transverse vibrations

    Torsional vibrations

4.2 EQUATION OF MOTION

There is two types of equation of motion

Longitudinal vibration of beam or axial vibration of a rod

Transverse vibration of a beam

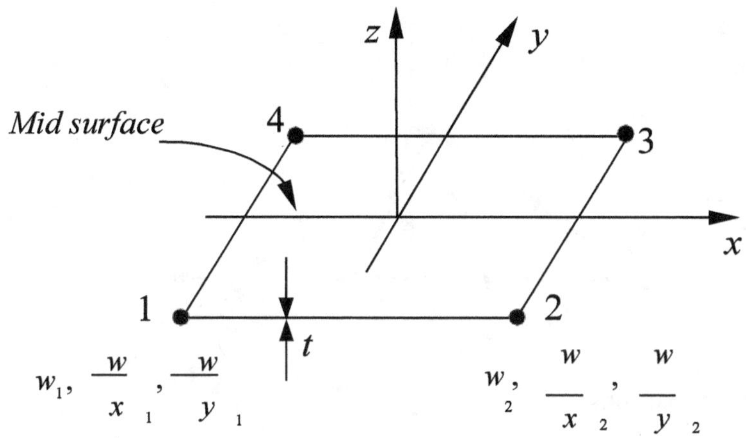

DOF at each node: $w,\ \mathrm{v}\ \frac{w}{y},\ \frac{w}{y}.$

On each element, the deflection *w(x,y)* is represented by

$$w(x, y) \qquad \underset{i\,1}{N_i w_i} \qquad N_{xi}\left(\frac{w}{x}\right)_i \qquad N_{yi}\left(\frac{w}{y}\right)_i ,$$

where N_i, N_{xi} and N_{yi} are shape functions. This is an incompatible element! The stiffness matrix is still of the form

$$\mathbf{k} = \quad \mathbf{B}^T\mathbf{E}\mathbf{B}dV,$$

where **B** is the strain-displacement matrix, and **E** the stress- strain matrix.

Minding Plate Elements:

4-Node Quadrilateral

8-Node Quadrilateral

Three independent fields.

Deflection $w(x,y)$ is linear for Q4, and quadratic for Q8.

Discrete Kirchhoff Element:

Triangular plate element (not available in *ANSYS*). Start with a 6-node riangular element,

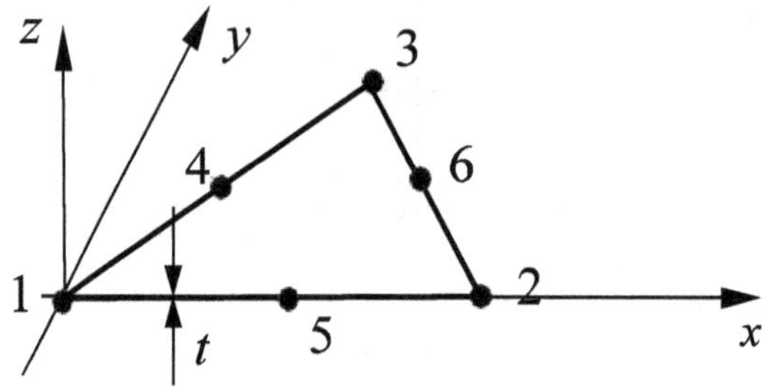

DOF at corner nodes: w, $\frac{w}{x}$, $\frac{w}{y}$, x, y

DOF at mid side nodes: Total DOF x, y.

= 21.

Then, impose conditions

xz yz 0, etc., at selected

nodes to reduce the DOF (using relations in (15)). Obtain:

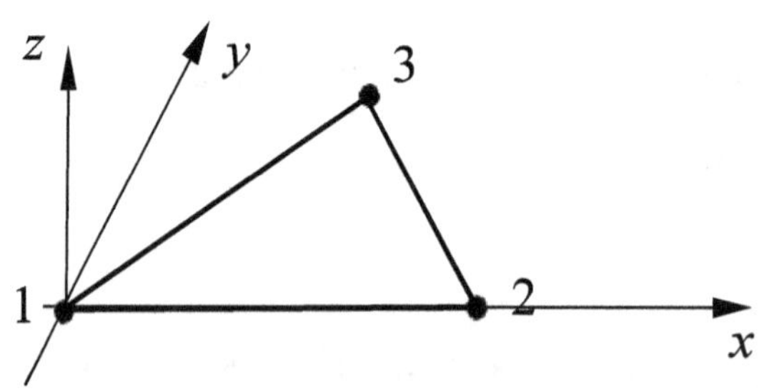

At each node: w, x $\frac{w}{x}$, y $\frac{w}{y}$.

Total DOF = 9 (DKT Element).

Incompatible $w(x,y)$; convergence is faster (w is cubic along each edge) and it is efficient.

Test Problem:

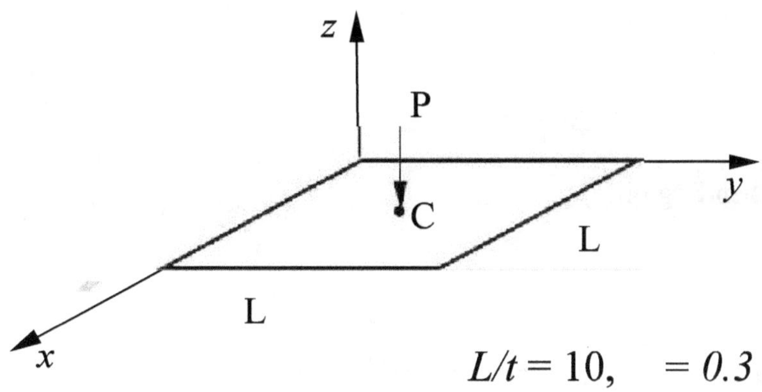

ANSYS 4-node quadrilateral plate element .

ANSYS Result for wc

Mesh	w_c (PL^2/D)
2 2	0.00593
4 4	0.00598
8 8	0.00574
16 16	0.00565
:	:
Exact Solution	0.00560

Question:Converges from "above"? Contradiction to what we learnt about the nature ofthe FEA solution?

Reason: This is an incompatible element (See comments on p. 177).

Shells and Shell Elements

Shells – Thin structures witch span over curved surfaces.

Example:

Sea shell, egg shell (the wonder of the nature); Containers, pipes, tanks;

Car bodies;

Roofs, buildings (the Superdome), etc.

Forces in shells:

Membrane forces + Bending Moments

(cf. plates: bending only)

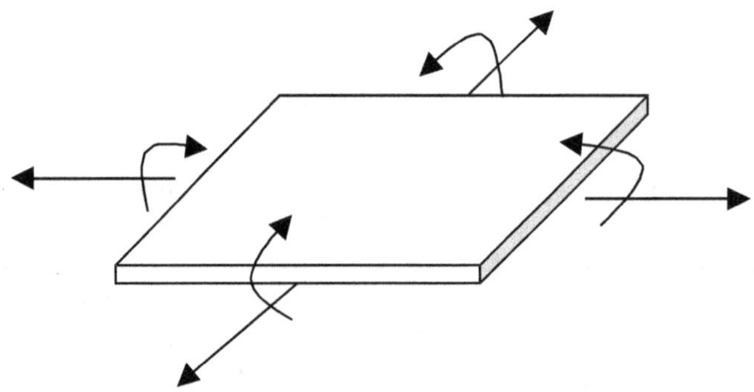

Example: A Cylindrical Container.

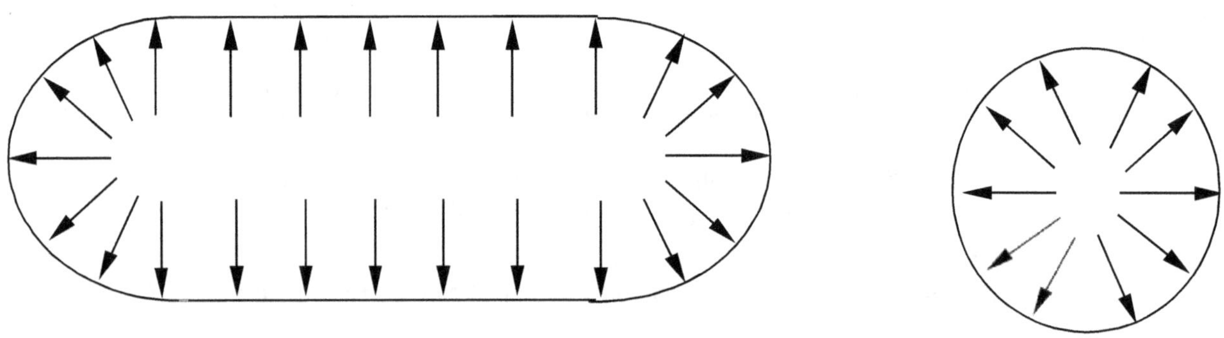

internal forces:

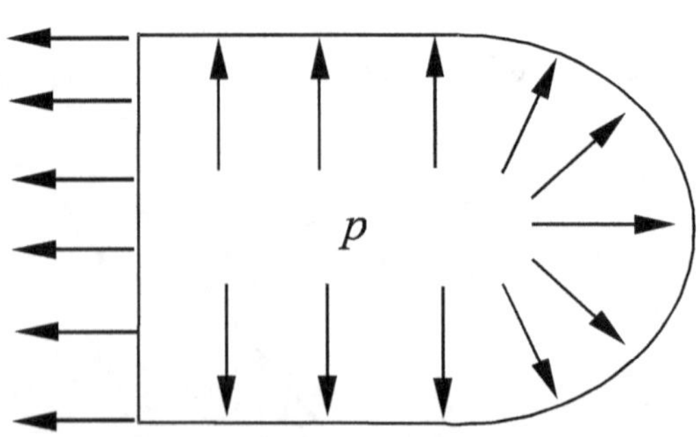

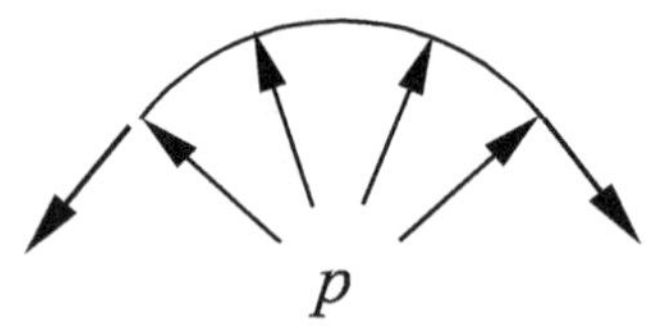

membrane stresses dominate

Shell Theory:

Thin shell theory

Shell theories are the most complicated ones to formulate and analyze in mechanics

(Russian's contributions).

Engineering   Craftsmanship Demand strong analytical skill

Shell Elements:

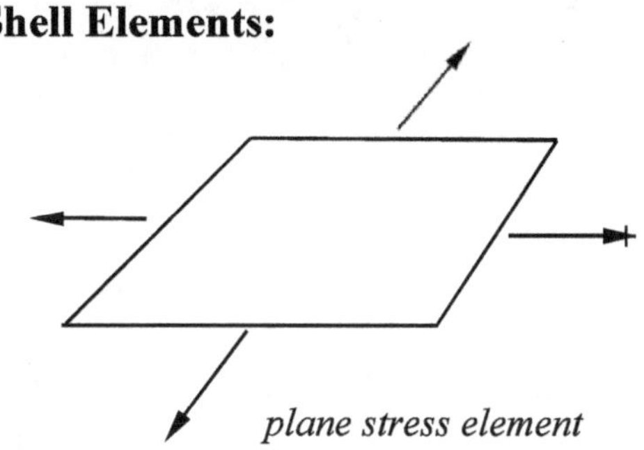

plane stress element

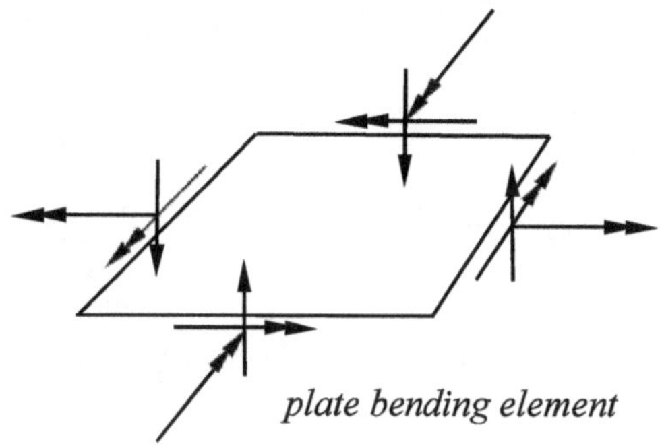

plate bending element

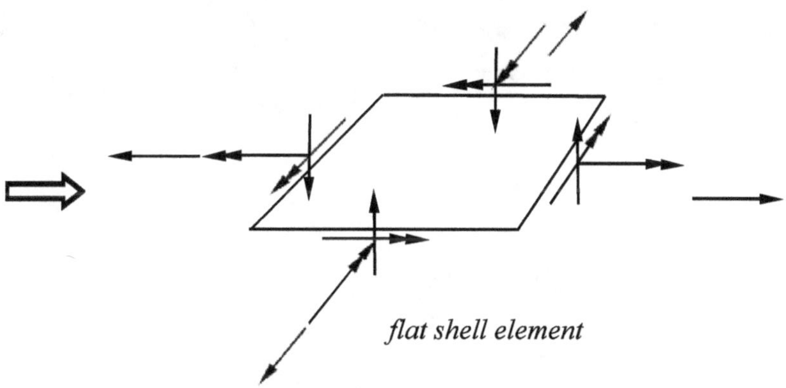

flat shell element

cf.: bar + simple beam element

=> general beam element.DOF

at each node:

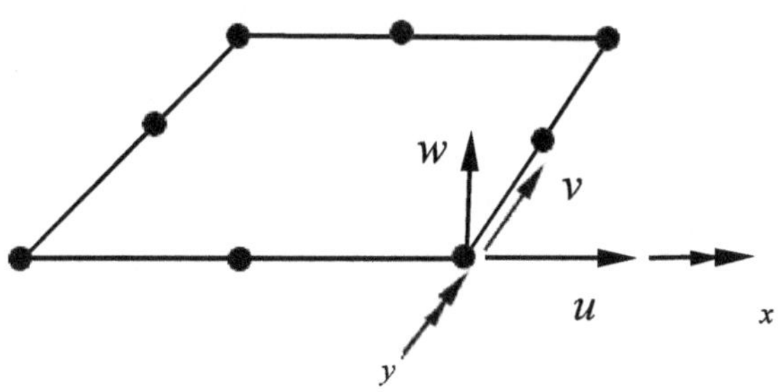

Q4 or Q8 shell element.

Curved shell elements:

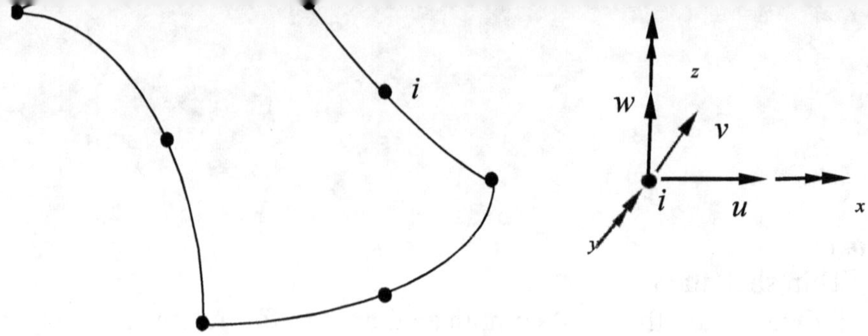
i
w
z
v
i
u
x
y

Based on shell theories;
Most general shell elements (flat shell and plate elements are subsets);Complicated in formulation.

Test Cases:

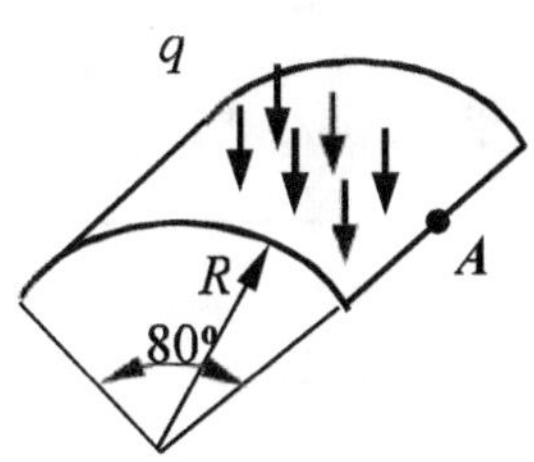

Roof

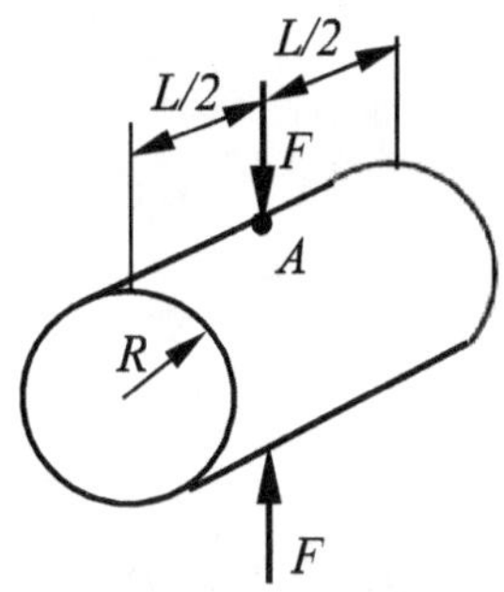

Pinched Cylinder

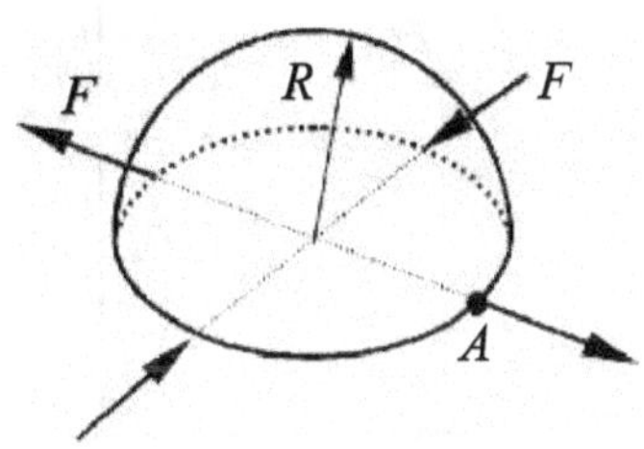

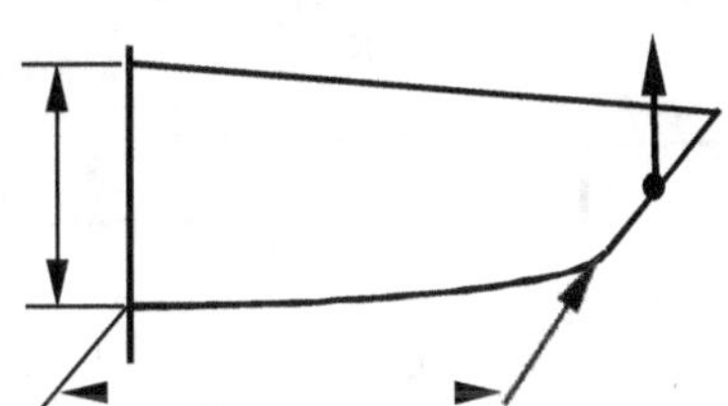

F

2

A

L

F_1

F

Pinched Hemisphere

4.3CONSISTENT MASS MATRICES

Natural frequencies and modes

Frequency

response

($F(t)=F_0 \sin wt$)

Transient

response ($F(t)$

arbitrary)

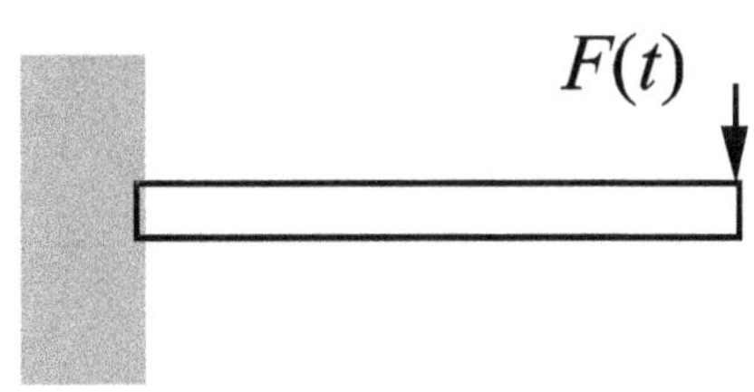

4.3.1 Single DOF System

4.3.2 Multiple DOF System

Reference SDOF Spring-Mass-Dashpot Model

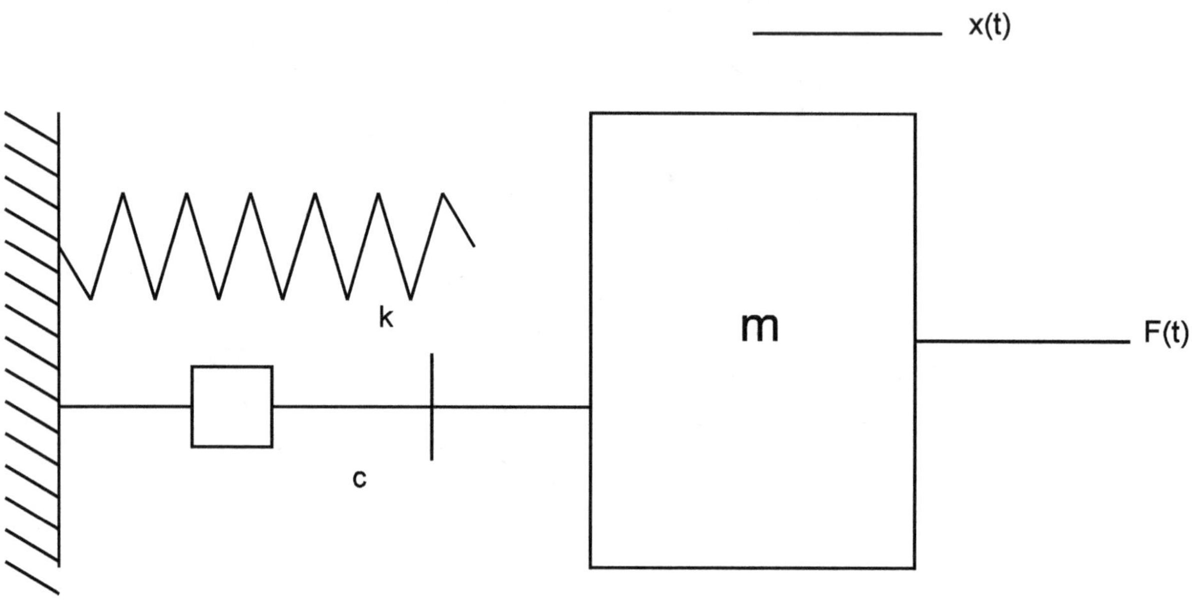

- Natural frequencies and modes
- Frequency response ($F(t)=F_o \sin\omega t$)
- Transient response ($F(t)$ arbitrary)

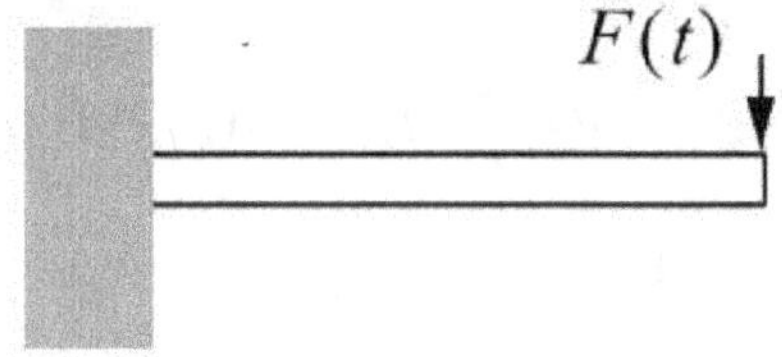

I. Basic Equations

A. Single DOF System

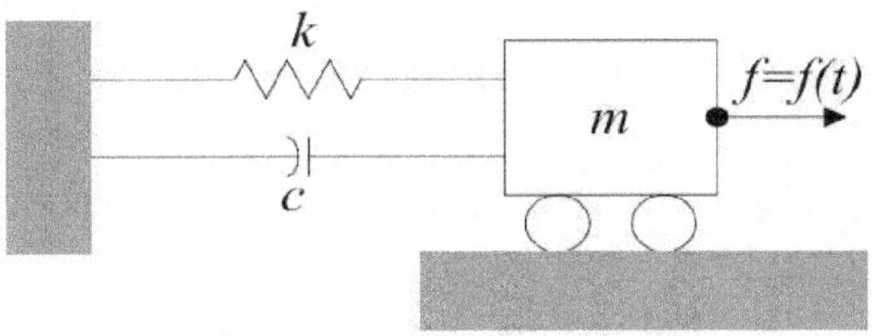

$$\begin{cases} m \text{ - mass} \\ k \text{ - stiffness} \\ c \text{ - damping} \\ f(t) \text{ - force} \end{cases}$$

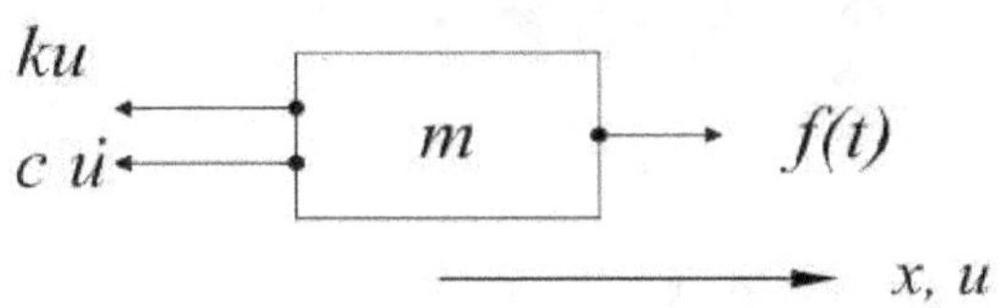

From Newton's law of motion ($ma = F$), we have

$$m\ddot{u} = f(t) - ku - c\dot{u},$$

i.e.

$$m\ddot{u} + c\dot{u} + ku = f(t), \qquad (1)$$

where u is the displacement, $\dot{u} = du/dt$ and $\ddot{u} = d^2u/dt^2$.

Free Vibration: $f(t) = 0$ and no damping ($c = 0$)

Eq. (1) becomes

$$m\ddot{u}+ku=0 \quad (2)$$

(meaning: inertia force + stiffness force = 0)

Assume:

$$u(t) = U \sin(\omega t),$$

where ω is the frequency of oscillation, U the amplitude.

Eq. (2) yields

$$-U\omega^2 m \sin(\omega t)+kU\sin(\omega t)=0$$

i.e.,

$$\left[-\omega^2 m+k\right]U=0.$$

For nontrivial solutions for U, we must have

$$\left[-\omega^2 m+k\right]=0,$$

which yields

$$\omega = \sqrt{\frac{k}{m}}. \quad (3)$$

This is the circular *natural frequency* of the single DOF system (rad/s). The cyclic frequency ($1/s$ = Hz) is

$$f=\frac{\omega}{2\pi}, \quad (4)$$

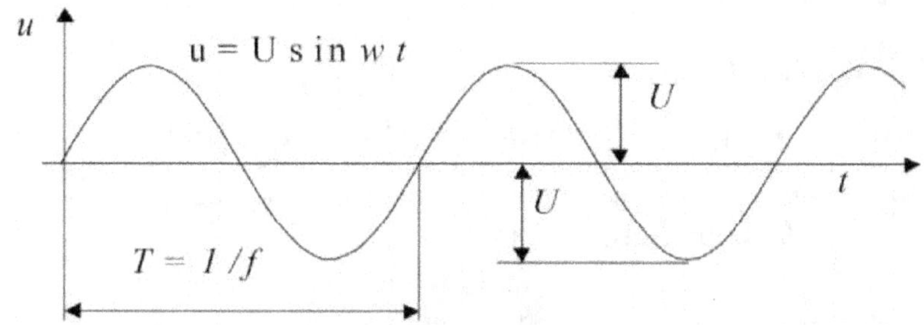

Undamped Free Vibration

With non-zero damping c, where

$$0 < c < c_c = 2m\omega = 2\sqrt{k\,m} \qquad (c_c = \text{critical damping}) \qquad (5)$$

we have the damped natural frequency:

$$\omega_d = \omega\sqrt{1-\xi^2}, \qquad (6)$$

where $\xi = \dfrac{c}{c_c}$ (damping ratio).

For structural damping: $0 \le \xi < 0.15$ (usually 1~5%)

$$\omega_d \approx \omega. \qquad (7)$$

Thus, we can ignore damping in normal mode analysis.

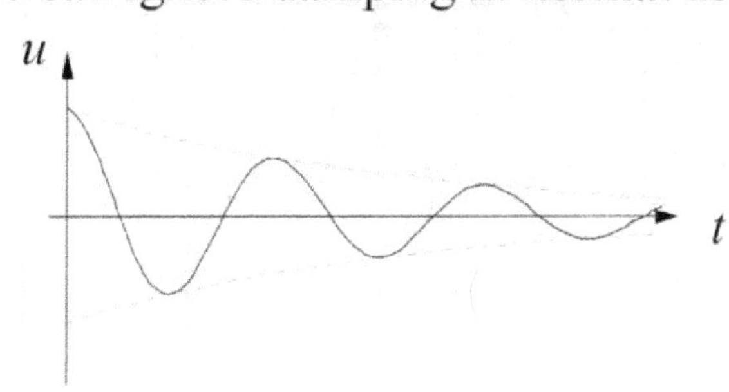

Damped Free Vibration

B. Multiple DOF System

Equation of Motion

Equation of motion for the whole structure is

$$\mathbf{M}\ddot{\mathbf{u}} + \mathbf{C}\dot{\mathbf{u}} + \mathbf{K}\mathbf{u} = \mathbf{f}(t),$$

in which: $\mathbf{u}$ — nodal displacement vector,

$\mathbf{M}$ — mass matrix,

$\mathbf{C}$ — damping matrix,

$\mathbf{K}$ — stiffness matrix,

$\mathbf{f}$ — forcing vector.

Physical meaning of Eq. (8):

Inertia forces + Damping forces + Elastic forces = Applied forces

Mass Matrices

Lumped mass matrix (1-D bar element):

$m_1 = \frac{\rho A L}{2}$

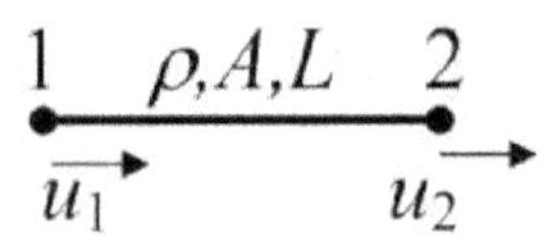

$m_2 = \frac{\rho A L}{2}$

Element mass matrix is found to be

$$\mathbf{m} = \underbrace{\begin{bmatrix} \frac{\rho A L}{2} & 0 \\ 0 & \frac{\rho A L}{2} \end{bmatrix}}_{\text{diagonal matrix}}$$

In general, we have the *consistent mass matrix* given by

$$\mathbf{m} = \int_V \rho \mathbf{N}^T \mathbf{N} dV \qquad (9)$$

where **N** is the same shape function matrix as used for the displacement field.

This is obtained by considering the kinetic energy:

$$\mathrm{K} = \frac{1}{2}\dot{\mathbf{u}}^T \mathbf{m}\dot{\mathbf{u}} \qquad (\text{cf. } \frac{1}{2}mv^2)$$

$$= \frac{1}{2}\int_V \rho \dot{u}^2 dV = \frac{1}{2}\int_V \rho (\dot{u})^T \dot{u} dV$$

$$= \frac{1}{2}\int_V \rho (\mathbf{N}\dot{\mathbf{u}})^T (\mathbf{N}\dot{\mathbf{u}}) dV$$

$$= \frac{1}{2}\dot{\mathbf{u}}^T \underbrace{\int_V \rho \mathbf{N}^T \mathbf{N} dV}_{\mathbf{m}} \dot{\mathbf{u}}$$

Bar Element (linear shape function):

$$\mathbf{m} = \int_V \rho \begin{bmatrix} 1-\xi \\ \xi \end{bmatrix} \begin{bmatrix} 1-\xi & \xi \end{bmatrix} AL d\xi$$

$$= \rho AL \begin{bmatrix} 1/3 & 1/6 \\ 1/6 & 1/3 \end{bmatrix} \begin{matrix} \ddot{u}_1 \\ \ddot{u}_2 \end{matrix} \qquad (10)$$

Element mass matrices:

⇒ local coordinates ⇒ to global coordinates

⇒ assembly of the global structure mass matrix **M**.

Example

Simple Beam Element:

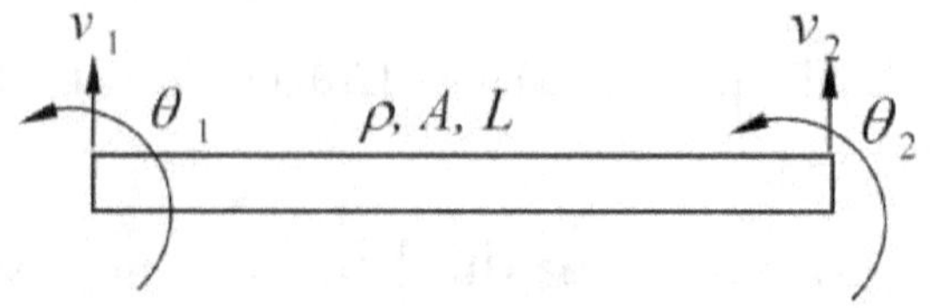

$$\mathbf{m} = \int_V \rho \mathbf{N}^T \mathbf{N} dV$$

$$= \frac{\rho A L}{420} \begin{bmatrix} 156 & 22L & 54 & -13L \\ 22L & 4L^2 & 13L & -3L^2 \\ 54 & 13L & 156 & -22L \\ -13L & -3L^2 & -22L & 4L^2 \end{bmatrix} \begin{matrix} \ddot{v}_1 \\ \ddot{\theta}_1 \\ \ddot{v}_2 \\ \ddot{\theta}_2 \end{matrix} \quad (11)$$

Units in dynamic analysis (make sure they are consistent):

	Choice I	***Choice II***
t (time)	s	s
L (length)	m	mm
m (mass)	kg	Mg
a (accel.)	m/s^2	mm/s^2
f (force)	N	N
ρ (density)	kg/m^3	Mg/mm^3

II. Free Vibration

Study of the dynamic characteristics of a structure:

- natural frequencies
- normal modes (shapes)

Let $\mathbf{f}(t) = \mathbf{0}$ and $\mathbf{C} = \mathbf{0}$ (ignore damping) in the dynamic equation (8) and obtain

$$\mathbf{M\ddot{u}} + \mathbf{Ku} = \mathbf{0} \tag{12}$$

Assume that displacements vary harmonically with time, that is,

$$\mathbf{u}(t) = \bar{\mathbf{u}} \sin(\omega t),$$
$$\dot{\mathbf{u}}(t) = \omega \bar{\mathbf{u}} \cos(\omega t),$$
$$\ddot{\mathbf{u}}(t) = -\omega^2 \bar{\mathbf{u}} \sin(\omega t),$$

where $\bar{\mathbf{u}}$ is the vector of nodal displacement amplitudes.

Eq. (12) yields,

$$\left[\mathbf{K} - \omega^2 \mathbf{M}\right]\bar{\mathbf{u}} = \mathbf{0} \tag{13}$$

This is a generalized eigenvalue problem (EVP).

Solutions?

Trivial solution: $\overline{\mathbf{u}} = \mathbf{0}$ for any values of ω (not interesting).

Nontrivial solutions: $\overline{\mathbf{u}} \neq \mathbf{0}$ only if

$$\left|\mathbf{K} - \omega^2 \mathbf{M}\right| = 0 \qquad (14)$$

This is an n-th order polynomial of ω^2, from which we can find n solutions (roots) or eigenvalues ω_i.

ω_i (i = 1, 2, …, n) are the natural frequencies (or characteristic frequencies) of the structure.

ω_1 (the smallest one) is called the fundamental frequency.

For each ω_i , Eq. (13) gives one solution (or eigen) vector

$$\left[\mathbf{K} - \omega_i^2 \mathbf{M}\right]\overline{\mathbf{u}}_i = \mathbf{0}.$$

$\overline{\mathbf{u}}_i$ (i=1,2,…,n) are the *normal modes* (or *natural modes*, *mode shapes*, etc.).

Properties of Normal Modes

$$\overline{\mathbf{u}}_i^T \mathbf{K} \overline{\mathbf{u}}_j = 0,$$

$$\overline{\mathbf{u}}_i^T \mathbf{M} \overline{\mathbf{u}}_j = 0, \qquad \text{for } i \neq j, \qquad (15)$$

if $\omega_i \neq \omega_j$. That is, modes are orthogonal (or independent) to each other with respect to **K** and **M** matrices.

Normalize the modes:

$$\bar{\mathbf{u}}_i^T \mathbf{M} \, \bar{\mathbf{u}}_i = 1,$$

$$\bar{\mathbf{u}}_i^T \mathbf{K} \, \bar{\mathbf{u}}_i = \omega_i^2 . \tag{16}$$

Note:

- Magnitudes of displacements (modes) or stresses in normal mode analysis have no physical meaning.
- For normal mode analysis, no support of the structure is necessary.

  $\omega_i = 0 \Leftrightarrow$ there are rigid body motions of the whole or a part of the structure.

  $\Rightarrow$ apply this to check the FEA model (check for mechanism or free elements in the models).
- Lower modes are more accurate than higher modes in the FE calculations (less spatial variations in the lower modes $\Rightarrow$ fewer elements/wave length are needed).

Example:

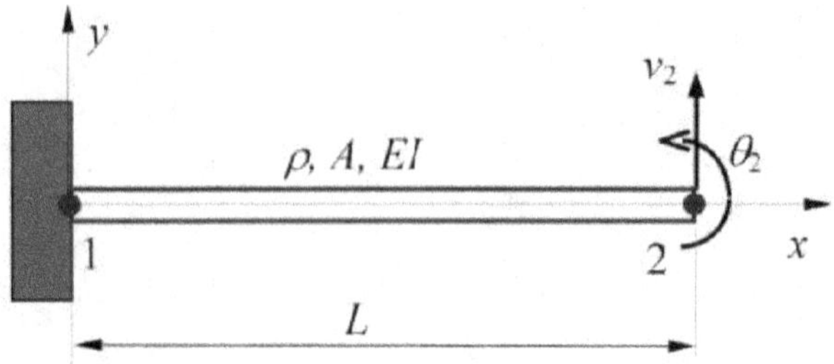

$$\left[\mathbf{K}-\omega^2\mathbf{M}\right]\begin{Bmatrix}\bar{v}_2\\ \bar{\theta}_2\end{Bmatrix}=\begin{Bmatrix}0\\0\end{Bmatrix},$$

$$\mathbf{K}=\frac{EI}{L^3}\begin{bmatrix}12 & -6L\\ -6L & 4L^2\end{bmatrix}, \qquad \mathbf{M}=\frac{\rho AL}{420}\begin{bmatrix}156 & -22L\\ -22L & 4L^2\end{bmatrix}.$$

EVP: $$\begin{vmatrix}12-156\lambda & -6L+22L\lambda\\ -6L+22L\lambda & 4L^2-4L^2\lambda\end{vmatrix}=0,$$

in which $\lambda=\omega^2\rho AL^4/420\,EI$.

Solving the EVP, we obtain,

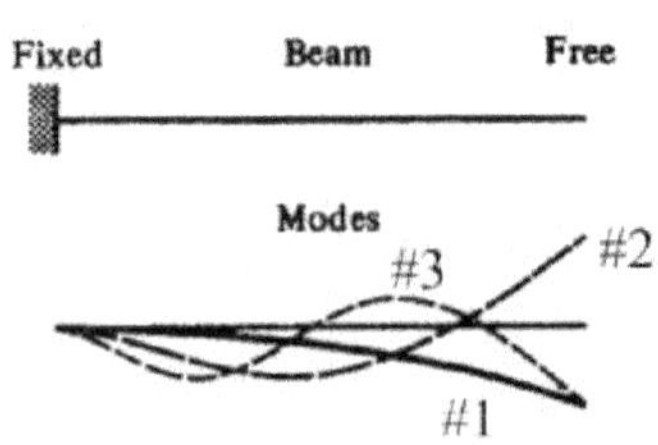

$$\omega_1=3.533\left(\frac{EI}{\rho AL^4}\right)^{1/2}, \quad \begin{Bmatrix}\bar{v}_2\\ \bar{\theta}_2\end{Bmatrix}_1=\begin{Bmatrix}1\\ 1.38/L\end{Bmatrix},$$

$$\omega_2=34.81\left(\frac{EI}{\rho AL^4}\right)^{1/2}, \quad \begin{Bmatrix}\bar{v}_2\\ \bar{\theta}_2\end{Bmatrix}_2=\begin{Bmatrix}1\\ 7.62/L\end{Bmatrix}.$$

Exact solutions:

$$\omega_1=3.516\left(\frac{EI}{\rho AL^4}\right)^{1/2}, \quad \omega_2=22.03\left(\frac{EI}{\rho AL^4}\right)^{1/2}.$$

We can see that mode 1 is calculated much more accurately than mode 2, with one beam element.

III. Damping

Two commonly used models for viscous damping.

A. Proportional Damping (Rayleigh Damping)

$$\mathbf{C} = \alpha\mathbf{M} + \beta\mathbf{K} \tag{17}$$

where the constants α & β are found from

$$\xi_1 = \frac{\alpha\omega_1}{2} + \frac{\beta}{2\omega_1}, \qquad \xi_2 = \frac{\alpha\omega_2}{2} + \frac{\beta}{2\omega_2},$$

with $\omega_1, \omega_2, \xi_1$ & ξ_2 (damping ratio) being selected.

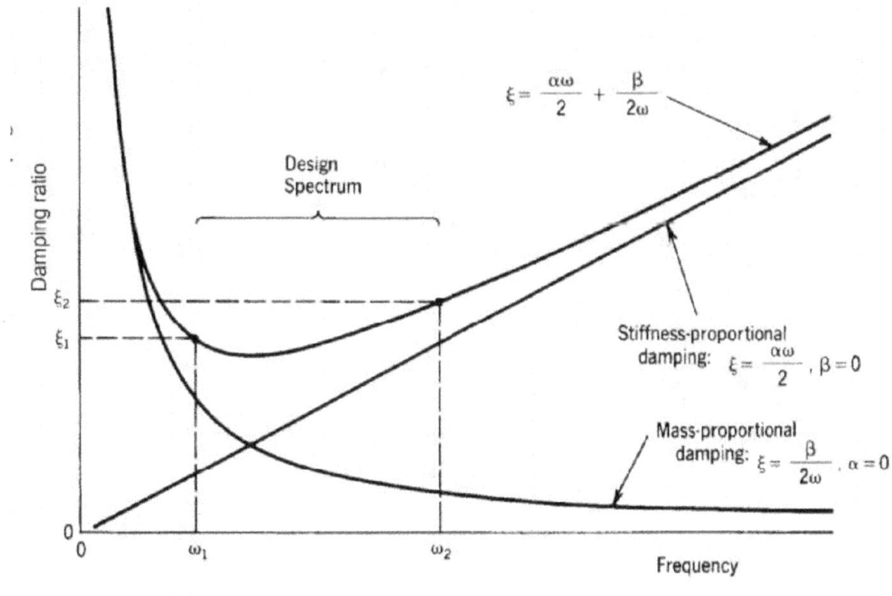

B. Modal Damping

Incorporate the viscous damping in modal equations.

IV. Modal Equations

Use the normal modes (modal matrix) to transform the coupled system of dynamic equations to uncoupled system of equations.

We have

$$\left[\boldsymbol{K} - \omega_i^2 \boldsymbol{M}\right]\bar{\boldsymbol{u}}_i = \boldsymbol{0}, \quad i = 1,2,\ldots, \text{n} \tag{18}$$

where the normal mode $\bar{\mathbf{u}}_i$ satisfies:

$$\begin{cases} \bar{\mathbf{u}}_i^T \mathbf{K} \bar{\mathbf{u}}_j = 0, \\ \bar{\mathbf{u}}_i^T \mathbf{M} \bar{\mathbf{u}}_j = 0, \end{cases} \quad \text{for i} \neq \text{j},$$

and

$$\begin{cases} \bar{\mathbf{u}}_i^T \mathbf{M} \bar{\mathbf{u}}_i = 1, \\ \bar{\mathbf{u}}_i^T \mathbf{K} \bar{\mathbf{u}}_i = \omega_i^2, \end{cases} \quad \text{for } i = 1, 2, \ldots, \text{n}.$$

Form the *modal matrix*:

$$\boldsymbol{\Phi}_{(n \times n)} = \left[\bar{\mathbf{u}}_1 \; \bar{\mathbf{u}}_2 \; \cdots \; \bar{\mathbf{u}}_n\right] \tag{19}$$

We can verify that

$$\boldsymbol{\Phi}^T \mathbf{K} \boldsymbol{\Phi} = \boldsymbol{\Omega} = \begin{bmatrix} \omega_1^2 & 0 & \cdots & 0 \\ 0 & \omega_2^2 & & \vdots \\ \vdots & & \ddots & 0 \\ 0 & \cdots & 0 & \omega_n^2 \end{bmatrix} \text{(Spectral matrix)}$$

$$\boldsymbol{\Phi}^T \mathbf{M} \boldsymbol{\Phi} = \mathbf{I}. \tag{20}$$

Transformation for the displacement vector,

$$\mathbf{u} = z_1 \bar{\mathbf{u}}_1 + z_2 \bar{\mathbf{u}}_2 + \cdots + z_n \bar{\mathbf{u}}_n = \boldsymbol{\Phi} \mathbf{z}, \tag{21}$$

where

$$\mathbf{z} = \begin{Bmatrix} z_1(t) \\ z_2(t) \\ \vdots \\ z_n(t) \end{Bmatrix}$$

are called *principal coordinates*.

Substitute (21) into the dynamic equation:

$$\mathbf{M}\,\Phi\,\ddot{\mathbf{z}} + \mathbf{C}\,\Phi\,\dot{\mathbf{z}} + \mathbf{K}\,\Phi\,\mathbf{z} = \mathbf{f}(t).$$

Pre-multiply by Φ^{T}, and apply (20):

$$\ddot{\mathbf{z}} + \mathbf{C}_{\phi}\dot{\mathbf{z}} + \Omega\,\mathbf{z} = \mathbf{p}(t), \qquad (22)$$

where $\mathbf{C}_{\phi} = \alpha\mathbf{I} + \beta\Omega$ (for proportional damping),

$$\mathbf{p} = \Phi^{T}\mathbf{f}(t).$$

Introduce *modal damping*:

$$\mathbf{C}_{\phi} = \begin{bmatrix} 2\xi_1\omega_1 & 0 & \cdots & 0 \\ 0 & 2\xi_2\omega_2 & & \\ \vdots & & \ddots & \vdots \\ 0 & & \cdots & 2\xi_n\omega_n \end{bmatrix}. \qquad (23)$$

Equation (22) becomes,

$$\ddot{z}_i + 2\xi_i\omega_i\dot{z}_i + \omega_i^2 z_i = p_i(t), \quad i = 1,2,\ldots,n. \quad (24)$$

Equations in (22) or (24) are called *modal equations*. These equations are uncoupled, second-order differential equations, which are much easier to solve than the original dynamic equation (a coupled system).

To recover **u** from **z,** apply transformation (21) again, once **z** is obtained from (24).

Notes:

- Only the first few modes may be needed in constructing the modal matrix Φ (i.e., Φ could be an n×m rectangular matrix with m<n). Thus, significant reduction in the size of the system can be achieved.
- Modal equations are best suited for problems in which higher modes are not important (i.e., structural vibrations, but not for structures under a shock load).

V. Frequency Response Analysis

(Harmonic Response Analysis)

$$\mathbf{M\ddot{u}} + \mathbf{C\dot{u}} + \mathbf{Ku} = \underbrace{\mathbf{F}\sin\omega t}_{\text{Harmonic loading}} \tag{25}$$

Modal method: Apply the modal equations,

$$\ddot{z}_i + 2\xi_i\omega_i\dot{z}_i + \omega_i^2 z_i = p_i \sin\omega t, \quad i=1,2,\ldots,m. \tag{26}$$

These are 1-D equations. Solutions are

$$z_i(t) = \frac{p_i/\omega_i^2}{\sqrt{(1-\eta_i^2)^2 + (2\xi_i\eta_i)^2}} \sin(\omega t - \theta_i), \tag{27}$$

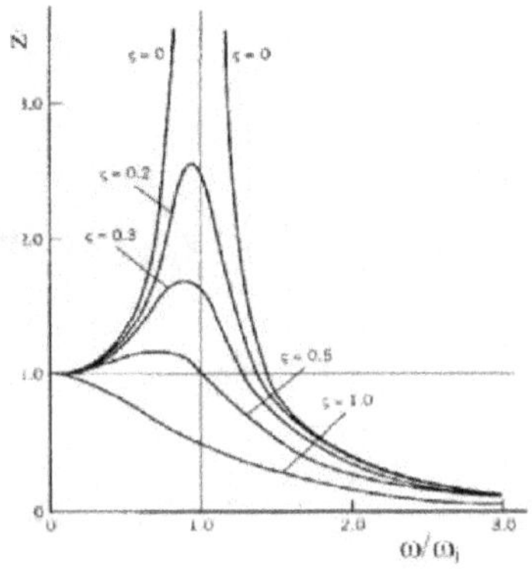

where

$$\begin{cases} \theta_i = \arctan \dfrac{2\xi_i\eta_i}{1-\eta_i^2}, \text{ phase angle} \\ \eta_i = \omega/\omega_i, \\ \xi_i = \dfrac{c_i}{c_c} = \dfrac{c_i}{2m\omega_i}, \text{ damping ratio} \end{cases}$$

Recover **u** from (21).

Direct Method: Solve Eq. (25) directly, that is, calculate the inverse. With $\mathbf{u} = \bar{\mathbf{u}} e^{i\omega t}$ (complex notation), Eq. (25) becomes

$$\left[\mathbf{K} + i\omega\mathbf{C} - \omega^2\mathbf{M}\right]\bar{\mathbf{u}} = \bar{\mathbf{F}}.$$

This equation is expensive to solve and matrix is ill-conditioned if ω is close to any ω_i.

VI. Transient Response Analysis

(Dynamic Response/Time-History Analysis)

- Structure response to *arbitrary, time-dependent loading.*

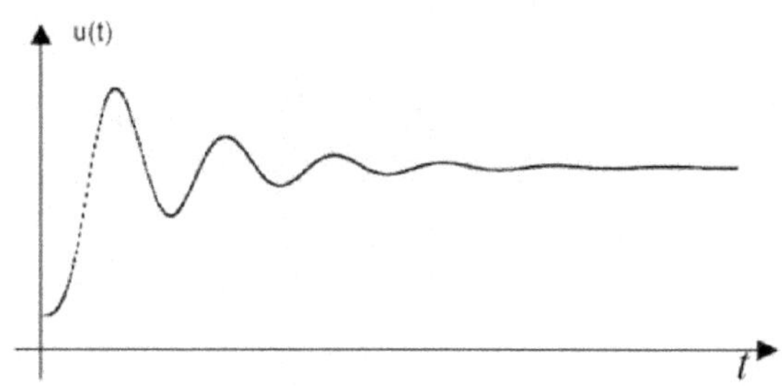

Compute responses by integrating through time:

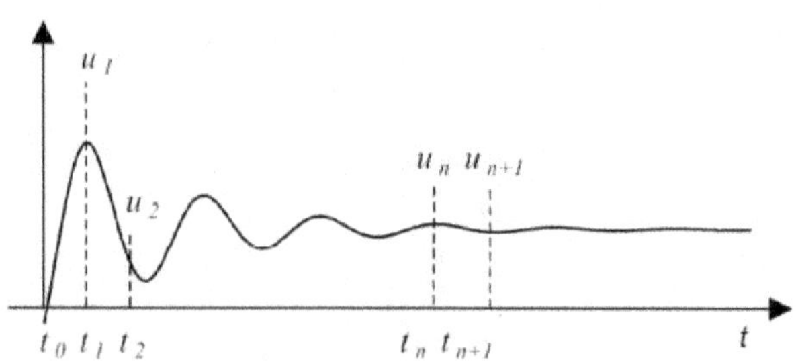

Equation of motion at instance t_n, $n = 0, 1, 2, 3, \cdots$:

$$\mathbf{M}\ddot{\mathbf{u}}_n + \mathbf{C}\dot{\mathbf{u}}_n + \mathbf{K}\mathbf{u}_n = \mathbf{f}_n.$$

Time increment: $\Delta t = t_{n+1} - t_n$, $n = 0, 1, 2, 3, \cdots$.

There are two categories of methods for transient analysis.

A. *Direct Methods (Direct Integration Methods)*

- ***Central Difference Method***

  Approximate using finite difference:

$$\dot{\mathbf{u}}_n = \frac{1}{2\Delta t}(\mathbf{u}_{n+1} - \mathbf{u}_{n-1}),$$

$$\ddot{\mathbf{u}}_n = \frac{1}{(\Delta t)^2}(\mathbf{u}_{n+1} - 2\mathbf{u}_n + \mathbf{u}_{n-1})$$

  Dynamic equation becomes,

$$\mathbf{M}\left[\frac{1}{(\Delta t)^2}(\mathbf{u}_{n+1} - 2\mathbf{u}_n + \mathbf{u}_{n-1})\right] + \mathbf{C}\left[\frac{1}{2\Delta t}(\mathbf{u}_{n+1} - \mathbf{u}_{n-1})\right] + \mathbf{K}\mathbf{u}_n = \mathbf{f}_n,$$

  which yields,

$$\mathbf{A}\mathbf{u}_{n+1} = \mathbf{F}(t)$$

  where

$$\begin{cases} \mathbf{A} = \dfrac{1}{(\Delta t)^2}\mathbf{M} + \dfrac{1}{2\Delta t}\mathbf{C}, \\ \mathbf{F}(t) = \mathbf{f}_n - \left[\mathbf{K} - \dfrac{2}{(\Delta t)^2}\mathbf{M}\right]\mathbf{u}_n - \left[\dfrac{1}{(\Delta t)^2}\mathbf{M} - \dfrac{1}{2\Delta t}\mathbf{C}\right]\mathbf{u}_{n-1}. \end{cases}$$

  $\mathbf{u}_{n+1}$ is calculated from $\mathbf{u}_n$ & $\mathbf{u}_{n-1}$, and solution is marching from $t_0, t_1, \cdots t_n, t_{n+1}, \cdots$, until convergent.

This method is *unstable* if Δt is too large.

- ***Newmark Method***:

Use approximations:

$$\mathbf{u}_{n+1} \approx \mathbf{u}_n + \Delta t \dot{\mathbf{u}}_n + \frac{(\Delta t)^2}{2}\left[(1-2\beta)\ddot{\mathbf{u}}_n + 2\beta \ddot{\mathbf{u}}_{n+1}\right] \rightarrow (\ddot{\mathbf{u}}_{n+1} = \cdots)$$

$$\dot{\mathbf{u}}_{n+1} \approx \dot{\mathbf{u}}_n + \Delta t\left[(1-\gamma)\ddot{\mathbf{u}}_n + \gamma \ddot{\mathbf{u}}_{n+1}\right],$$

where β & γ are chosen constants. These lead to

$$\mathbf{A}\mathbf{u}_{n+1} = \mathbf{F}(t)$$

where

$$\mathbf{A} = \mathbf{K} + \frac{\gamma}{\beta \Delta t}\mathbf{C} + \frac{1}{\beta (\Delta t)^2}\mathbf{M},$$

$$\mathbf{F}(t) = f(\mathbf{f}_{n+1}, \gamma, \beta, \Delta t, \mathbf{C}, \mathbf{M}, \mathbf{u}_n, \dot{\mathbf{u}}_n, \ddot{\mathbf{u}}_n).$$

This method is *unconditionally stable* if

$$2\beta \geq \gamma \geq \frac{1}{2}.$$

$$\text{e.g.,} \quad \gamma = \frac{1}{2}, \; \beta = \frac{1}{4}$$

which gives the constant average acceleration method.

Direct methods can be expensive! (the need to compute $\mathbf{A}^{-1}$, often repeatedly for each time step).

B. Modal Method

First, do the transformation of the dynamic equations using the modal matrix before the time marching:

$$\mathbf{u} = \sum_{i=1}^{m} \bar{\mathbf{u}}_i z_i(t) = \Phi\, \mathbf{z}, \qquad i = 1, 2, \cdots, m.$$
$$\ddot{z}_i + 2\xi_i \omega_i \dot{z}_i + \omega_i z_i = p_i(t),$$

Then, solve the uncoupled equations using an integration method. Can use, e.g., 10%, of the total modes (m= n/10).

- Uncoupled system,
- Fewer equations,
- No inverse of matrices,
- More efficient for large problems.

Comparisons of the Methods

Direct Methods	*Modal Method*
• Small model	• Large model
• More accurate (with small Δt)	• Higher modes ignored
• Single loading	• Multiple loading
• Shock loading	• Periodic loading
• …	• …

Two objectives:

- Determine the temperature field (steady or unsteady state)
- Stresses due to the temperature changes

I. Temperature Field

Fourier Heat Conduction Equation:

1-D Case:

$$f_x = -k\frac{\partial T}{\partial x}, \tag{1}$$

where,

f_x = heat flux per unit area,
k = thermal conductivity,
$T = T(x)$ = temperature.

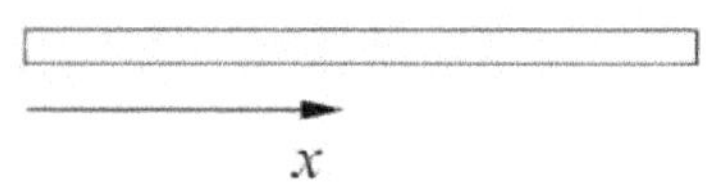

3-D Case:

$$\begin{Bmatrix} f_x \\ f_y \\ f_z \end{Bmatrix} = -\mathbf{K}\begin{Bmatrix} \partial T/\partial x \\ \partial T/\partial y \\ \partial T/\partial z \end{Bmatrix}, \tag{2}$$

where, f_x, f_y, f_z = heat flux in x, y and z direction, respectively, and in case of isotropy,

$$\mathbf{K} = \begin{bmatrix} k & 0 & 0 \\ 0 & k & 0 \\ 0 & 0 & k \end{bmatrix}. \tag{3}$$

The Equation of Heat Flow is:

$$-\left[\frac{\partial f_x}{\partial x}+\frac{\partial f_y}{\partial y}+\frac{\partial f_z}{\partial z}\right]+q_v = c\rho\frac{\partial T}{\partial t} \tag{4}$$

in which,

q_v = rate of internal heat generation per unit volume,

c = specific heat,

ρ = mass density.

For steady state ($\partial T/\partial t = 0$) and isotropic materials, we can obtain:

$$k\nabla^2 T = -q_v. \tag{5}$$

This a Poisson equation.

Boundary Conditions (BC's):

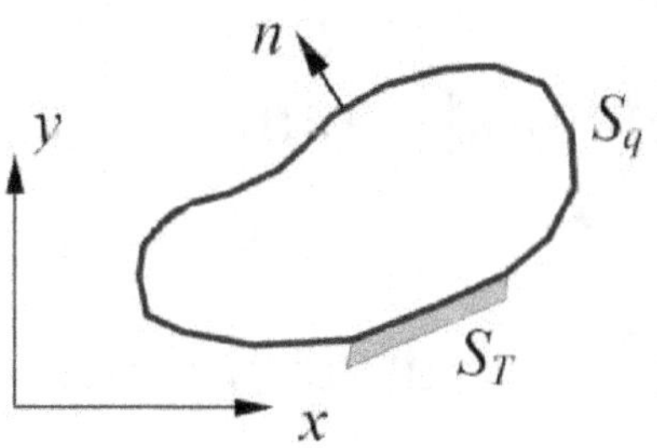

$$T = \overline{T}, \qquad \text{on } S_T;$$

$$\frac{\partial T}{\partial n} = \overline{Q}, \qquad \text{on } S_q. \tag{6}$$

Note that at any point on the boundary $S = S_T \cup S_q$, only one type of BC can be specified.

Finite Element Formulation for Heat Conduction:

$$\mathbf{K}_T \mathbf{T} = \mathbf{q} \quad (7)$$

where,

$\mathbf{K}_T$ = conductivity matrix,

$\mathbf{T}$ = vector of nodal temperature,

$\mathbf{q}$ = vector of thermal loads.

The element conductivity matrix is given by:

$$\mathbf{k}_T = \int_V \mathbf{B}^T \mathbf{K} \mathbf{B} dV . \quad (8)$$

This is obtained in a similar way as for the structural analysis, e.g., by starting with the interpolation $T = \mathbf{N}\mathbf{T}_e$ ($\mathbf{N}$ is the shape function matrix, T_e the nodal temperature).

Note that there is only one DOF at each node for the thermal problems.

Thermal Transient Analysis:

$$\frac{\partial T}{\partial t} \neq 0 .$$

Apply FDM (use time steps and integrate in time), as in the transient structural analysis, to obtain the transient temperature fields.

II. Thermal Stress Analysis

- Solve Eq. (7) first to obtain the temperature (change) fields.
- Apply the temperature change ΔT as initial strains (or initial stresses) to the structure.

1-D Case:

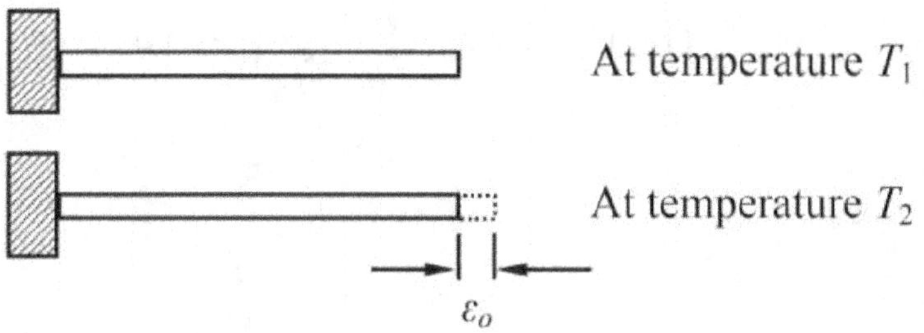

Thermal Strain (Initial Strain):

$$\varepsilon_o = \alpha \Delta T, \tag{9}$$

in which,

α = the coefficient of thermal expansion,

$\Delta T = T_2 - T_1$ is the change of temperature.

Total strain,

$$\varepsilon = \varepsilon_e + \varepsilon_o \tag{10}$$

with ε_e being the elastic strain due to mechanical load.

That is,

$$\varepsilon = E^{-1}\sigma + \alpha \Delta T, \tag{11}$$

or $$\sigma = E(\varepsilon - \varepsilon_o). \tag{12}$$

Example: The above shown bar under thermal load ΔT.

(a) If no constraint on the right-hand side, that is, the bar is free to expand to the right, then

$$\varepsilon = \varepsilon_o, \quad \varepsilon_e = 0, \quad \sigma = 0,$$

from Eq. (12). No thermal stress!

(b) If there is a constraint on the right-hand side, that is, the bar can not expand to the right, then

$$\varepsilon = 0, \quad \varepsilon_e = -\varepsilon_o = -\alpha\Delta T, \quad \sigma = -E\alpha\Delta T,$$

from Eqs. (10) and (12). Thus, thermal stress exists!

2-D Cases:

Plane Stress,

$$\boldsymbol{\varepsilon}_o = \begin{Bmatrix} \varepsilon_x \\ \varepsilon_y \\ \gamma_{xy} \end{Bmatrix}_o = \begin{Bmatrix} \alpha\Delta T \\ \alpha\Delta T \\ 0 \end{Bmatrix}. \tag{13}$$

Plane Strain,

$$\varepsilon_o = \begin{Bmatrix} \varepsilon_x \\ \varepsilon_y \\ \gamma_{xy} \end{Bmatrix}_o = \begin{Bmatrix} (1+\nu)\alpha\Delta T \\ (1+\nu)\alpha\Delta T \\ 0 \end{Bmatrix}. \tag{14}$$

Here, ν is the Poisson's ratio.

3-D Case:

$$\boldsymbol{\varepsilon}_o = \begin{Bmatrix} \varepsilon_x \\ \varepsilon_y \\ \varepsilon_z \\ \gamma_{xy} \\ \gamma_{yz} \\ \gamma_{zx} \end{Bmatrix}_o = \begin{Bmatrix} \alpha\Delta T \\ \alpha\Delta T \\ \alpha\Delta T \\ 0 \\ 0 \\ 0 \end{Bmatrix}. \tag{15}$$

Observation: Temperature changes do not yield shear strains.

Total Strain:

$$\boldsymbol{\varepsilon} = \boldsymbol{\varepsilon}_e + \boldsymbol{\varepsilon}_o. \tag{16}$$

Stress-Strain Relation:

$$\boldsymbol{\sigma} = \mathbf{E}\boldsymbol{\varepsilon}_e = \mathbf{E}(\boldsymbol{\varepsilon} - \boldsymbol{\varepsilon}_o). \tag{17}$$

Thermal Stress Analysis Using the FEM:

- Need to specify α for the structure and ΔT on the related elements (which experience the temperature change).
- Note that for linear thermoelasticity, same temperature change will yield same stresses, even if the structure is at two different temperatures.
- Differences in the temperatures during the manufacturing and working environment are the main cause of thermal (residual) stresses.

4.6 TRANSIENT RESPONSE ANALYSIS

(Dynamic Response/Time-History Analysis)

Structure response to *arbitrary, time-dependent loading.*

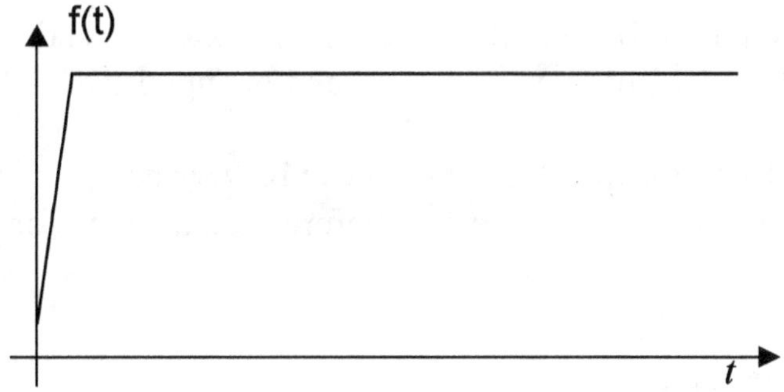

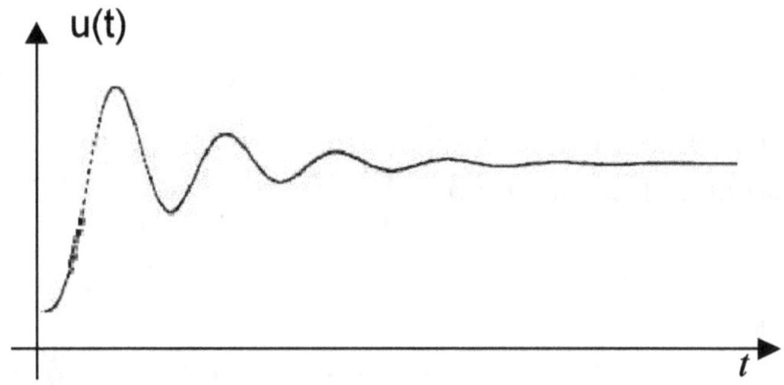

Compute responses by integrating through time:

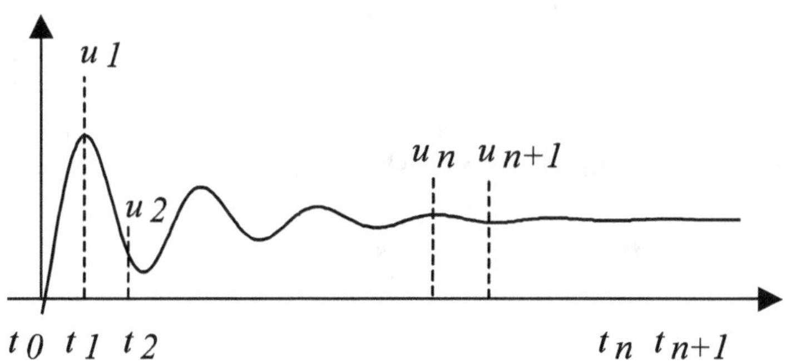

B. Modal Method

First, do the transformation of the dynamic equations using the modal matrix before the time marching:

Then, solve the uncoupled equations using an integration method. Can use, e.g.,10%, of the total modes (m= n/10).

Uncoupled system, Fewer equations,

No inverse of matrices,
More efficient for large problems.

4.6.1 Cautions in Dynamic Analysis

Symmetry: It should not be used in the dynamic analysis (normal modes, etc.) because symmetric structures can have antisymmetric modes.

Mechanism, rigid body motion means = 0. Can use this to check FEA models to see if they are properly connected and/or supported.

Input for FEA: loading *F(t)* or *F( )* can be very complicated in real applications and often needs to be filtered first before used as input for FEA.

Examples

Impact, drop test, etc.

PROBLEM

In the spring structure shown $k_1 = 10$ lb./in., $k_2 = 15$ lb./in., $k_3 = 20$ lb./in., P= 5 lb. Determine the deflection at nodes 2 and 3.

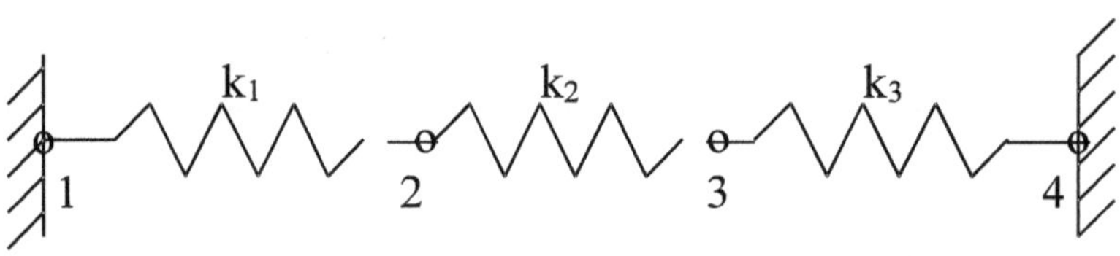

Figure 2.4

Solution:

Again apply the three steps outlined previously.

Step 1: Find the Element Stiffness Equations

Element 1:

$$[K^{(1)}] = \begin{pmatrix} 10 & -10 \\ -10 & 10 \end{pmatrix} \begin{matrix} 1 \\ 2 \end{matrix} \quad (\text{columns } 1, 2)$$

Element 2:

$$[K^{(2)}] = \begin{pmatrix} 15 & -15 \\ -15 & 15 \end{pmatrix} \begin{matrix} 2 \\ 3 \end{matrix} \quad (\text{columns } 2, 3)$$

Element 3:

$$[K^{(3)}] = \begin{pmatrix} 20 & -20 \\ -20 & 20 \end{pmatrix} \begin{matrix} 3 \\ 4 \end{matrix} \quad (\text{columns } 3, 4)$$

Step 2: Find the Global stiffness matrix

$$\begin{array}{c} \\ 1 \\ 2 \\ 3 \\ 4 \end{array}\begin{bmatrix} 1 & 2 & 3 & 4 \\ 10 & -10 & 0 & 0 \\ -10 & 10+15 & -15 & 0 \\ 0 & -15 & 15+20 & -20 \\ 0 & 0 & -20 & 20 \end{bmatrix} = \begin{bmatrix} 10 & -10 & 0 & 0 \\ -10 & 25 & -15 & 0 \\ 0 & -15 & 35 & -20 \\ 0 & 0 & -20 & 20 \end{bmatrix}$$

Now the global structural equation can be written as,

$$\begin{Bmatrix} F_1 \\ F_2 \\ F_3 \\ F_4 \end{Bmatrix} = \begin{bmatrix} 10 & -10 & 0 & 0 \\ -10 & 25 & -15 & 0 \\ 0 & -15 & 35 & -20 \\ 0 & 0 & -20 & 20 \end{bmatrix} \begin{Bmatrix} u_1 \\ u_2 \\ u_3 \\ u_4 \end{Bmatrix}$$

Step 3: Solve for Deflections

The known boundary conditions are: $u_1 = u_4 = 0$, $F_3 = P = 3$lb. Thus, rows and columns 1 and 4 will drop out, resulting in the following matrix equation,

$$\begin{Bmatrix} 0 \\ 3 \end{Bmatrix} = \begin{bmatrix} 25 & -15 \\ -15 & 35 \end{bmatrix} \begin{Bmatrix} u_2 \\ u_3 \end{Bmatrix}$$

Solving, we get $u_2 = 0.0692$ & $u_3 = 0.1154$

PROBLEM

In the spring structure shown, $k_1 = 10$ N/mm, $k_2 = 15$ N/mm, $k_3 = 20$ N/mm, $k_4 = 25$ N/mm, $k_5 = 30$ N/mm, $k_6 = 35$ N/mm. $F_2 = 100$ N. Find the deflections in all springs.

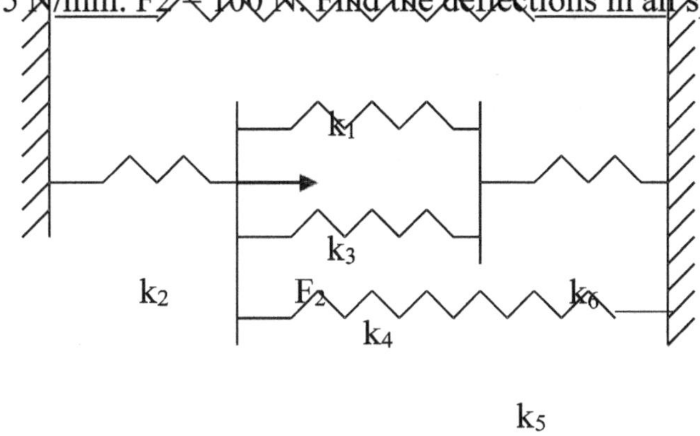

Solution:

Here again, we follow the three-step approach described earlier, without specifically mentioning at each step.

Element 1:

$$[K^{(1)}] = \begin{array}{c} \begin{matrix} \;\;1 & \;\;\;4 \end{matrix} \\ \begin{pmatrix} 10 & -10 \\ -10 & 10 \end{pmatrix} \end{array} \begin{matrix} \\ 1 \\ 4 \end{matrix}$$

Element 2:

$$[K^{(2)}] = \begin{array}{c} \begin{matrix} \;\;1 & \;\;\;2 \end{matrix} \\ \begin{pmatrix} 15 & -15 \\ -15 & 15 \end{pmatrix} \end{array} \begin{matrix} \\ 1 \\ 2 \end{matrix}$$

Element 3:

$$[K^{(3)}] = \begin{array}{c} \begin{matrix} \;\;2 & \;\;\;3 \end{matrix} \\ \begin{pmatrix} 20 & -20 \\ -20 & 20 \end{pmatrix} \end{array} \begin{matrix} \\ 2 \\ 3 \end{matrix}$$

Element 4:

$$[K^{(4)}] = \begin{array}{c} \begin{matrix} \;\;2 & \;\;\;3 \end{matrix} \\ \begin{pmatrix} 25 & -25 \\ -25 & 25 \end{pmatrix} \end{array} \begin{matrix} \\ 2 \\ 3 \end{matrix}$$

Element 5:

$$[K^{(5)}] = \begin{array}{c} \begin{matrix} \;\;2 & \;\;\;4 \end{matrix} \\ \begin{pmatrix} 30 & -30 \\ -30 & 30 \end{pmatrix} \end{array} \begin{matrix} \\ 2 \\ 4 \end{matrix}$$

Element 6:

$$[K^{(6)}] = \begin{array}{c} \begin{matrix} \;\;3 & \;\;\;4 \end{matrix} \\ \begin{pmatrix} 35 & -35 \\ -35 & 35 \end{pmatrix} \end{array} \begin{matrix} \\ 3 \\ 4 \end{matrix}$$

The global stiffness matrix is,

$$[K_g] = \begin{array}{c} \begin{matrix} 1 & 2 & 3 & 4 \end{matrix} \\ \begin{pmatrix} 10+15 & -15 & 0 & -10 \\ -15 & 15+20+25+30 & -20-25 & -30 \\ 0 & -20-25 & 20+25+35 & -35 \\ -10 & -30 & -35 & 10+30+35 \end{pmatrix} \end{array} \begin{matrix} \\ 1 \\ 2 \\ 3 \\ 4 \end{matrix}$$

And simplifying, we get

$$[K_g] = \begin{pmatrix} 25 & -15 & 0 & -10 \\ -15 & 90 & -45 & -30 \\ 0 & -45 & 80 & -35 \\ -10 & -30 & -35 & 75 \end{pmatrix}$$

And the structural equation is,

$$\begin{Bmatrix} F_1 \\ F_2 \\ F_3 \\ F_4 \end{Bmatrix} = \begin{pmatrix} 25 & -15 & 0 & -10 \\ -15 & 90 & -45 & -30 \\ 0 & -45 & 80 & -35 \\ -10 & -30 & -35 & 75 \end{pmatrix} \begin{Bmatrix} u_1 \\ u_2 \\ u_3 \\ u_4 \end{Bmatrix}$$

Now, apply the boundary conditions, $u_1 = u_4 = 0$, $F_2 = 100$ N. This is carried out by deleting the rows 1 and 4, columns 1 and 4, and replacing F_2 by 100N. The final matrix equation is,

Which $\begin{Bmatrix} 100 \\ 0 \end{Bmatrix} = \begin{pmatrix} 90 & -45 \\ -45 & 80 \end{pmatrix} \begin{Bmatrix} u_2 \\ u_3 \end{Bmatrix}$ gives

Deflections:

Spring 1: $u_4 - u_1 = 0$

Spring 2: $u_2 - u_1 = 1.54590$

Spring 3: $u_3 - u_2 = -0.6763$

Spring 4: $u_3 - u_2 = -0.6763$

Spring 5: $u_4 - u_2 = -1.5459$

Spring 6: $u_4 - u_3 = -0.8696$

5.3 SCALE VARIABLE PROBLEM IN 2 DIMENSIONS

$$u = \begin{Bmatrix} x \\ y \end{Bmatrix} = \begin{bmatrix} N_1 & 0 & N_2 & 0 & N_3 & 0 & N_4 & 0 \\ 0 & N_1 & 0 & N_2 & 0 & N_3 & 0 & N_4 \end{bmatrix} \begin{Bmatrix} x_1 \\ y_2 \\ x_1 \\ y_2 \\ x_3 \\ y_3 \\ x_4 \\ y_4 \end{Bmatrix}$$

$N_1=1/4(1-\varepsilon)(1-\eta)$; $N_2=1/4(1+\varepsilon)(1-\eta)$; $N_3=1/4(1+\varepsilon)(1+\eta)$; $N_4=1/4(1-\varepsilon)(1+\eta)$.

- **Equation of Stiffness Matrix for 4 noded isoparametric quadrilateral element**

$$[J]=\begin{bmatrix} J_{11} & J_{12} \\ J_{21} & J_{22} \end{bmatrix};$$

$$[B]=\frac{1}{|J|}\begin{bmatrix} J_{22} & -J_{12} & 0 & 0 \\ 0 & 0 & -J_{21} & J_{11} \\ -J_{21} & J_{11} & J_{22} & -J_{12} \end{bmatrix}\begin{bmatrix} \frac{\partial N_1}{\partial \varepsilon} & 0 & \frac{\partial N_2}{\partial \varepsilon} & 0 & \frac{\partial N_3}{\partial \varepsilon} & 0 & \frac{\partial N_4}{\partial \varepsilon} & 0 \\ \frac{\partial N_1}{\partial \eta} & 0 & \frac{\partial N_2}{\partial \eta} & 0 & \frac{\partial N_3}{\partial \eta} & 0 & \frac{\partial N_4}{\partial \eta} & 0 \\ 0 & \frac{\partial N_1}{\partial \varepsilon} & 0 & \frac{\partial N_2}{\partial \varepsilon} & 0 & \frac{\partial N_3}{\partial \varepsilon} & 0 & \frac{\partial N_4}{\partial \varepsilon} \\ 0 & \frac{\partial N_1}{\partial \eta} & 0 & \frac{\partial N_2}{\partial \eta} & 0 & \frac{\partial N_3}{\partial \eta} & 0 & \frac{\partial N_4}{\partial \eta} \end{bmatrix}$$

$$[D]=\frac{E}{(1-\nu^2)}\begin{bmatrix} 1-\nu & \nu & 0 \\ \nu & 1 & 0 \\ 0 & 0 & \frac{1-\nu}{2} \end{bmatrix}$$, for plane stress conditions;

$$[D]=\frac{E}{(1+\nu)(1-2\nu)}\begin{bmatrix} 1-\nu & \nu & 0 \\ \nu & 1-\nu & 0 \\ 0 & 0 & \frac{1-2\nu}{2} \end{bmatrix}$$, for plane strain conditions.

- **Equation of element force vector**

$$\{F\}_e=[N]^T\begin{Bmatrix} F_x \\ F_y \end{Bmatrix};$$

N – Shape function, F_x – load or force along x direction,

F_y – load or force along y direction.

- **Numerical Integration (Gaussian Quadrature)**

The Gauss quadrature is one of the numerical integration methods to calculate the definite integrals. In FEA, this Gauss quadrature method is mostly preferred. In this method the numerical integration is achieved by the following expression,

$$\int_{-1}^{1} f(x)dx=\sum_{i=1}^{n} w_i f(x_i)$$

Table gives gauss points for integration from -1 to 1.

Number of Points n	Location x_i	Corresponding Weights w_i
1	$x_1 = 0.000$	**2.000**
2	$x_1, x_2 = \pm\sqrt{\frac{1}{3}} = \pm 0.577350269189$	**1.000**
3	$x_1, x_3 = \pm\sqrt{\frac{3}{5}} = \pm 0.774596669241$ $x_2 = 0.000$	$\frac{5}{9} = 0.555555$ $\frac{8}{9} = 0.888888$
4	$x_1, x_4 = \pm 0.8611363116$ $x_2, x_3 = \pm 0.3399810436$	**0.3478548451** **0.6521451549**

- **Problem (I set)**

1. Evaluate $I = \int_{-1}^{1} \cos \frac{\pi x}{2} dx$, by applying 3 point Gaussian quadrature and compare with exact solution.
2. Evaluate $I = \int_{-1}^{1} \left[3e^x + x^2 + \frac{1}{x+2} \right] dx$, using one point and two point Gaussian quadrature. Compare with exact solution.
3. For the isoparametric quadrilateral element shown in figure, determine the local co –ordinates of the point P which has Cartesian co-ordinates (7, 4).

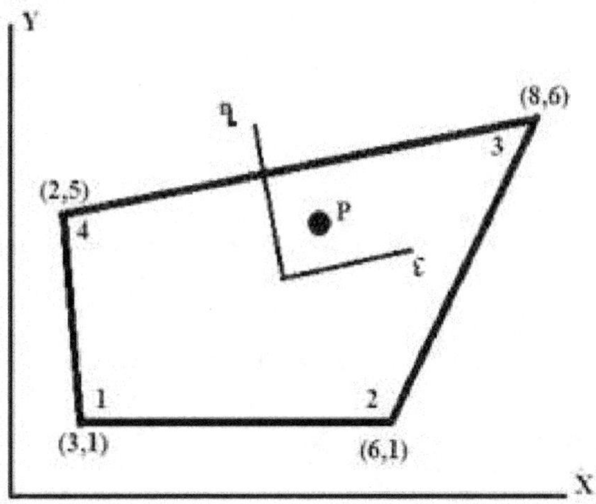

4. A four noded rectangular element is in figure. Determine (i) Jacobian matrix, (ii) Strain – Displacement matrix and (iii) Element Stresses. Take $E=2x10^5 N/mm^2$, $\upsilon = 0.25$, $u=[0,0,0.003,0.004,0.006, 0.004,0,0]^T$, $\varepsilon = 0$, $\eta=0$. Assume plane stress condition.

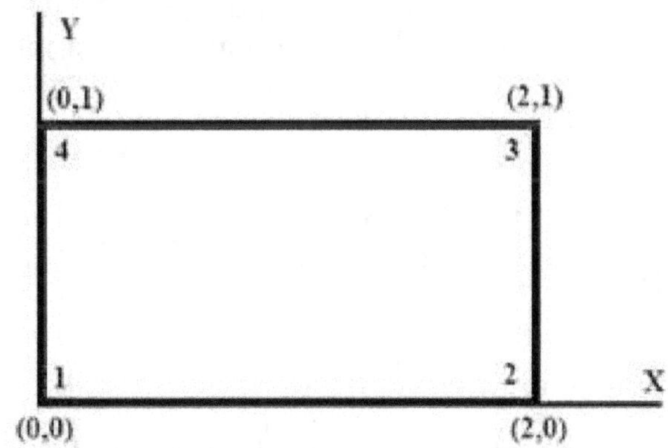

5.4 2 DIMENSIONAL FLUID MECHANICS

The problem of linear elastostatics described in detail in can be extended to include the effects of inertia. The resulting equations of motion take the form

$$\begin{aligned} \nabla \cdot \sigma + f &= \rho \ddot{u} && \text{in } \Omega \times I, \\ \sigma n &= \bar{t} && \text{on } \Gamma_q \times I, \\ u &= \bar{u} && \text{on } \Gamma_u \times I, \\ u(x1, x2, x3, 0) &= u_0(x1, x2, x3) && \text{in } \Omega, \\ v(x1, x2, x3, 0) &= v_0(x1, x2, x3) && \text{in } \Omega, \end{aligned}$$

Fluid Mechanics

The finite elementmethodsapplied to two-dimensionalfluid dynamic problemsis presented in this chapter. It is shown that even for the most basic transport equation numerical difficulties arise when the traditional finite element method is used to solve hyperbolic equations. Moreover,an iterative solver,required to handlethe nonlinear convectiveterms inherently presentin the inviscid Euler equations,is described.

Finite elements in fluid dynamics

As mentionedin the finite elementmethod (FEM) hasbeenoriginally proposed as a tool to solvestructural or solid mechanicsproblemsspecially in the aircraft industry. The method proved to be so powerful that it quickly spreadto diverseareassuch as heat conduction or potential flow simulations. The adequacyof the FEM is strongly related to the mathematical nature of the problem being solved. Wheneverthe governingpartial differential equations are elliptic [23], or at least parabolic, the accuracy of the method hasbeenvalidated time and again despiteknown difficulties suchas shearlocking in plate bending problems. Elliptic or parabolic partial differential equations can be shown to be associatedto quadratic functionals whoseminimization lead to the original governing equations.In thesesituations the Galerkin formulation of the finite elementmethod leadsto symmetric stiffness matrices and the difference between exact and approximate solutions is minimized with respect to an energy norm [13, 24]The Laplace equation in heat conduction or the fourth order differential equation in plate bending problems are examples of elliptic equations.

The tremendous success of the finite element method in structural problems motivated its use in fluid dynamics problems.It was expected that its good accuracy and nice convergence properties would be extended to fluid dynamic problems.These expectation were, however,unfounded. It was soon observed that numericalinstabilities appeared,even for the most basic transport equation, whenever the convection terms were dominant , terms result in non-symmetric matrices and the best approximation property prevailingin structural problems is no longer valid. The numerical instabilities can be alleviated bymesh refinement but this solution rapidly leads to problems so large that they cannot besolved in a reasonable time frame.

Therefore, if the finite element method is be used as a practical tool to model fluid dynamics problems, then there must be a way around the numerical instabilities other than mesh refinement.In fact, the traditional Galerkin approach can be modified such that stabilization terms are added to the governing equations such that these terms become zero when the problem is dominated by diffusive mechanism and are otherwise active, i.e., when the problem is dominated by convective mechanism.These stabilization schemes were a key to the utilization of the finite element method in fluid dynamics.

Lagrangian and Eulerian descriptions

Problems in solid mechanics are described in the Lagrangian or material coordinate system whereas problems in fluid dynamics are described in the Eulerian or spatial coordinate system. In the material coordinate system the interest lies in following the path of the particles composing the continuum. On the other hand, in the spatial coordinate system the interest lies in a fixed region in the space through which the continuum passes.

The Lagrangiancoordinatesystemmay seemmore intuitive becauseit tracks the movement of all the particles of interest. Indeed, it is simpler to write the equationsof motion and other conservationlaws is terms of material coordinateswhat makesit the most used description strategy when it comesto structural mechanics.Wheneverit is possibleto assign a referenceframe to a systemof particles that do not presentseveredistortions during its movementthen the Lagrangiancoordinatesystemis certainly the bestoption. However, in fluid mechanics,it is not uncommonto observeflows that presentseveredistortions due to turbulence or shock waves,what makesthe Eulerian description more suited. The disadvantagesof the Eulerian description are: the inherent presenceof nonlinear convective terms in the governingequationsand its clumsinessto handle free surfaceand interface flow problems.

The material coordinates$\bar{x}$ will be identified by an overbarwhereasthe spatial coordinatesare denotedsimply by x. Figure 3.1 showsa fixed bundleof particles at three instants of time: $C(t_0)$, $C(t_1)$ and $C(t_2)$. The material coordinates$\bar{x}$ refer to any given particle of that bundle. Notice that the spatial coordinatesx are not attached to any givenparticle of the bundle; it refersto any given point in space.In fact, one could imagine that, between instants of time t_1 and t_2, someof the particles in the bundle havepassedthrough point P definedby vector x.

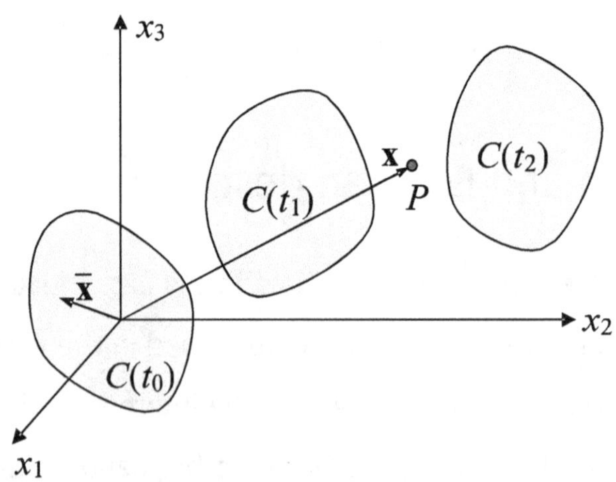

Figure 3.1: Material and spatial coordinate systems

A distinction can be made betweentotal and updated Lagrangiandescriptions. In the total Lagrangian description the material coordinatesx refer to a fixed configuration at someinstant of time, usually the initial time. In the updated Lagrangian description the referenceconfiguration changes foreachinstant of time and, therefore, must be changed as time progresses. The updated Lagrangian is usually employed in large deformation problemswhere the referencesystemmust be updated in betweentime or load steps.

It is possibleto introduce a relationship betweenthe material and the spatial coordinates suchthat

$$x = x(\underline{x}, t) \tag{3.1}$$

and to use that relationship to describethe continuum properties with respect to one or another coordinate system. In order for a one-to-one correspondence to exist, the transformation needsthat $\det(\partial x/\partial \underline{x}) > 0$. This one-to-onecorrespondenceallows one to trace back the history of the particle becausean inverserelationship is well-defined.

Consider that a continuum property ϕ has to be described sometimes in terms of the spatial coordinates and sometimes in terms of the material coordinate systemThen, the following equality can be written:

$$\bar{\phi}(\underline{x}, t) = \phi(x, t), \tag{3.2}$$

where the symbols ϕ and $\bar{\phi}$ have been purposedly used to highlight that they represent different function forms. Differentiation of equation (3.2) with respect to space yields

$$\frac{\partial \bar{\phi}}{\partial \underline{x}} = \frac{\partial \phi}{\partial x}\frac{\partial x}{\partial \underline{x}} \tag{3.3}$$

and, since $\det(\partial x/\partial \underline{x}) > 0$, there is no difficulty in relating the gradients in the spatial and material coordinate systems.Differentiation of equation (3.2) with respect to time yields

$$\frac{\partial \bar{\phi}}{\partial t} = \frac{\partial \phi}{\partial t} + \frac{\partial \phi}{\partial x} \cdot \frac{\partial x}{\partial t} = \frac{\partial \phi}{\partial t} + v \cdot \frac{\partial \phi}{\partial x} = \frac{\partial \phi}{\partial t} + v \cdot \nabla \phi, \tag{3.4}$$

where $v = \partial x/\partial t$ is the fluid velocity. Equation (3.4) may be differently expressedas

$$\left.\frac{\partial \phi}{\partial t}\right|_{X} = \left.\frac{\partial \phi}{\partial t}\right|_{x} + v \cdot \nabla\phi \quad \text{or} \quad \frac{d\phi}{dt} = \frac{\partial \phi}{\partial t} + v \cdot \nabla\phi. \tag{3.5}$$

The operator $d(\cdot)/dt$ in equation (3.5) is the material time derivative and the operator $\partial(\cdot)/\partial t$ is the spatial time derivative.

Conservation equations

The conservation laws involve rate of variation of the continuum p roperties with time. For example, Newton's second law,or conservation of linear momentum,relates time rate of change of the linear momentum with external forces .Therefore, it is important to establishrelationships that may render easier the task of obtaining the mathematical expression of the conservation laws.

Consider a bundle of particles Ω at time t in space bounded by a smooth surface Γ as shown in figure 3.2a where n is the unit normal on surface Γ and v is the fluid velocity introduced in equation (3.4). At time $t + \Delta t$ these particles have moved to and they occupy a new region of space $\Omega_{t+\Delta t}$. The instantaneous time rate of change of a property Φ associated with the bundle of particles is

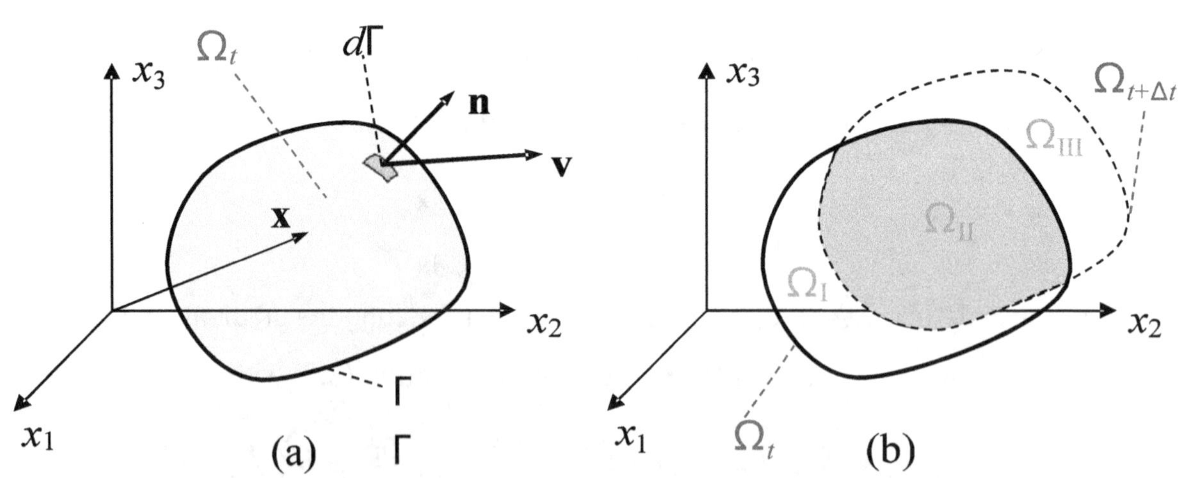

Figure 3.2: Transport theorem sketch

$$\frac{d\Phi}{dt} = \lim_{\Delta t \to 0} \frac{1}{\Delta t} \left[\int_{\Omega_{t+\Delta t}} \phi(x, t + \Delta t) d\Omega - \int_{\Omega_t} \phi(x, t) d\Omega \right], \tag{3.6}$$

where ϕ is the the value of Φ per unit mass.

The difficulty in computing equation (3.6) is that the region of integration changes from Ω_t at time t to $\Omega_{t+\Delta t}$ at time $t + \Delta t$. In order to facilitate the computations consider three regions as shown in figure 3.2b designated by Ω_I, Ω_{II} and Ω_{III} such that $\Omega_t = \Omega_I \cup \Omega_{II}$ and $\Omega_{t+\Delta t} = \Omega_{II} \cup \Omega_{III}$. Equation (3.7) can be written as

$$\begin{aligned}
\frac{d\Phi}{dt} &= \lim_{\Delta t \to 0} \frac{1}{\Delta t} \left[\int_{\Omega_{II} \cup \Omega_{III}} \phi(x, t + \Delta t) d\Omega - \int_{\Omega_I \cup \Omega_{II}} \phi(x, t) d\Omega \right] \\
&= \lim_{\Delta t \to 0} \frac{1}{\Delta t} \left[\int_{\Omega_{II}} \phi(x, t + \Delta t) d\Omega + \int_{\Omega_{III}} \phi(x, t + \Delta t) d\Omega - \int_{\Omega_I} \phi(x, t) d\Omega - \int_{\Omega_{II}} \phi(x, t) d\Omega \right] \\
&= \lim_{\Delta t \to 0} \frac{1}{\Delta t} \left[\int_{\Omega_I} \phi(x, t + \Delta t) d\Omega + \int_{\Omega_{II}} \phi(x, t + \Delta t) d\Omega + \int_{\Omega_{III}} \phi(x, t + \Delta t) d\Omega - \int_{\Omega_I} \phi(x, t + \Delta t) d\Omega - \int_{\Omega_I} \phi(x, t) d\Omega - \int_{\Omega_{II}} \phi(x, t) d\Omega \right] \\
&= \lim_{\Delta t \to 0} \frac{1}{\Delta t} \left[\int_{\Omega_t} \phi(x, t + \Delta t) d\Omega - \int_{\Omega_t} \phi(x, t) d\Omega \right] + \lim_{\Delta t \to 0} \frac{1}{\Delta t} \left[\int_{\Omega_{III}} \phi(x, t + \Delta t) d\Omega - \int_{\Omega_I} \phi(x, t + \Delta t) d\Omega \right] \\
&= \int_{\Omega_t} \frac{\partial \phi}{\partial t} d\Omega + \lim_{\Delta t \to 0} \frac{1}{\Delta t} \left[\int_{\Omega_{III}} \phi(x, t + \Delta t) d\Omega - \int_{\Omega_I} \phi(x, t + \Delta t) d\Omega \right].
\end{aligned} \tag{3.7}$$

The integrals over regions Ω_I and Ω_{III} in the last line of equation (3.7) represent the flux of ϕ through the boundary Γ of Ω_t. When $\Delta t \to 0$ those volume integrals can be expressed as a surface integral as in equation (3.8):

$$\frac{d\Phi}{dt} = \frac{d}{dt} \int_\Omega \phi(x, t) d\Omega = \underbrace{\int_\Omega \frac{\partial \phi(x, t)}{\partial t} d\Omega}_{\text{spatial rate of change}} + \underbrace{\int_\Gamma \phi(x, t) v \cdot n d\Gamma}_{\text{flux through } \Gamma}. \tag{3.8}$$

Equation (3.8) is known as the Reynolds transport theorem and states that the time rate of change of the property $\phi(x, t)$ over Ω is composed of two terms: the spatial rate of change and the transport of property ϕ through Γ.

Mass equation

Mass can be neither created nor destroyed in classical mechanics. Therefore, assuming continuum density ρ, the Reynolds transport theorem for $\phi = \rho$ reads:

$$0 = \frac{d}{dt}\int_{\Omega} \rho \, d\Omega = \int_{\Omega} \frac{\partial \rho}{\partial t} d\Omega + \int_{\Gamma} \rho \mathbf{v} \cdot \mathbf{n} \, d\Gamma = \int_{\Omega} \left[\frac{\partial \rho}{\partial t} + \nabla \cdot (\rho \mathbf{v}) \right] d\Omega, \tag{3.9}$$

where the Gauss or divergence theorem has been employed. Since equation (3.9) holds for an arbitrary control volume

$$\frac{\partial \rho}{\partial t} + \nabla \cdot (\rho \mathbf{v}) = 0. \tag{3.10}$$

Equation (3.10) is the mass conservation equation that can be re-written using the abridged form $\partial(\cdot)/\partial\alpha = (\cdot)_{,\alpha}$ and a tensorial notation of repeated indices:

$$\rho_{,t} + (\rho v_j)_{,x_j} = 0. \tag{3.11}$$

Linear momentum equation

Newton's second law states that the rate of change of linear momentum equals the total external forces.The linear momentum is a vector quantity defined as $\rho \mathbf{v}$Hence, using the transport equation (3.8),

$$\frac{d}{dt}\int_{\Omega} \rho \mathbf{v} \, d\Omega = \int_{\Omega} \left[\frac{\partial \rho \mathbf{v}}{\partial t} + \nabla \cdot (\rho \mathbf{v}\mathbf{v}) \right] d\Omega. \tag{3.12}$$

The quantity $\mathbf{v}\mathbf{v}$ present in equation (3.12) is in fact a second order tensorl.n order to simplify it the corresponding tensorial notation can be used to yield

$$\begin{aligned} \frac{\partial \rho \mathbf{v}}{\partial t} + \nabla \cdot (\rho \mathbf{v}\mathbf{v}) &= (\rho v_i e_i)_{,t} + (\rho v_i e_i v_j)_{,x_j} \\ &= [\rho_{,t} + (\rho v_j)_{,x_j}] v_i e_i + \rho (v_i e_i)_{,t} + \rho v_j (v_i e_i)_{,x_j} \\ &= \rho[(v_i e_i)_{,t} + v_j (v_i e_i)_{,x_j}] = \rho \left[\frac{\partial \mathbf{v}}{\partial t} + \mathbf{v} \cdot (\nabla \mathbf{v}) \right] = \rho \frac{d\mathbf{v}}{dt}, \end{aligned} \tag{3.13}$$

where equation (3.11) has been used, e_i is the unit vector in the direction of the x_i axis and v_i is the componentof v in axis x_i.

The external forcesacting are of two kinds: (i) a boundary force t with normal and tangential components that relate to the stress tensor σ according to $t = \sigma \cdot n$ and (ii) a body force per unit mass b. Therefore, substitution of equation (3.13) into (3.12) leads to the linear momentum conservation equation:

$$\frac{d}{dt}\int_\Omega \rho v d\Omega = \int_\Omega \rho \frac{dv}{dt} d\Omega = \int_\Omega \rho b d\Omega + \int_\Gamma \sigma \cdot n d\Gamma = \int_\Omega [\rho b + \nabla \cdot \sigma] d\Omega \quad (3.14)$$

and, since Ω is arbitrary,

$$\rho \frac{dv}{dt} = \rho b + \nabla \cdot \sigma \text{ or } \frac{\partial(\rho v)}{\partial t} + \nabla \cdot (\rho v v) = \rho b + \nabla \cdot \sigma. \quad (3.15)$$

Energy equation

The energy equation is the mathematical expression of the first law of thermodynamics that states that energy can be neither destroyed nor created although it may change from one form to the other. Making e the energy per unit mass the energy conservation equations is

$$\frac{d}{dt}\int_\Omega \rho e d\Omega = Q + \dot{W}, \quad (3.16)$$

where Q is the heat added to the particles within Ω and $\dot{W}$ is the rate of work done on the particles. Notice that equation (3.16) is in fact a power conservation law instead of an energy conservation law.However,it ultimately causes no difficulties. The heat added is given by

$$Q = -\int_\Gamma q \cdot n d\Gamma, \quad (3.17)$$

where q is the heat flux vector.The minus sign is present because Q is positive when heat flows into the domain Ω.

The rate of work doneby the external boundary force $t = \sigma \cdot n$ and the body force b is

$$\dot{W} = \int_\Omega \rho v \cdot b d\Omega + \int_\Gamma v \cdot t d\Gamma. \tag{3.18}$$

Substitution of equations(3.17) and (3.18) into (3.16) yields

$$\begin{aligned} \frac{d}{dt}\int_\Omega \rho e d\Omega &= -\int_\Gamma q \cdot n d\Gamma + \int_\Omega \rho v \cdot b d\Omega + \int_\Gamma v \cdot t d\Gamma \\ &= -\int_\Omega \nabla \cdot q d\Omega + \int_\Omega \rho v \cdot b d\Omega + \int_\Gamma v \cdot (\sigma \cdot n) d\Gamma \\ &= \int_\Omega [-\nabla \cdot q + \rho v \cdot b + \nabla \cdot (v \cdot \sigma)] d\Omega. \end{aligned} \tag{3.19}$$

Sincethe volume Ω is arbitrary, useof the transport theorem equation (3.8) yields

$$\frac{\partial(\rho e)}{\partial t} + \nabla \cdot (\rho e v) = -\nabla \cdot q + \rho v \cdot b + \nabla \cdot (v \cdot \sigma). \tag{3.20}$$

Euler equations

Euler equations are a specialform of the conservation equations (3.10),(3.15) and (3.20) for the case when the fluid is inviscid and the flow adiabatic.In order to derive the Euler equations it is important to take a closer look at the structure of the stress tensor σ. Figure 3.3 shows the stress tensor components in cartesian coordinates.

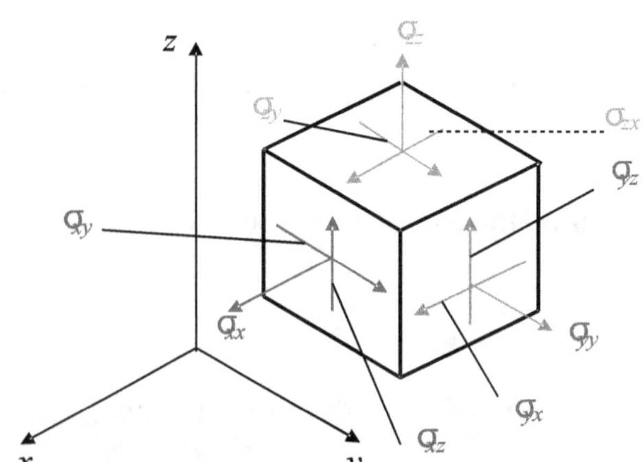

Figure 3.3: Stress tensor components

It is customary to split the stresstensor σ into two parts suchthat

$$\sigma = -pI + \tau, \tag{3.21}$$

where I is the unit tensor, p is the pressureand τ is the viscousstresstensor.

If the flow is adiabatic then $q = 0$. If it is inviscid $\tau = 0$. Under theseconditions, and recalling equation (3.21), equations (3.10), (3.15) and (3.20) can be written as

$$\begin{aligned}
&\frac{\partial \rho}{\partial t} + \nabla \cdot (\rho v) = 0 \\
&\frac{\partial (\rho v)}{\partial t} + \nabla \cdot (\rho v v) + \nabla p = \rho b \\
&\frac{\partial (\rho e)}{\partial t} + \nabla \cdot ((\rho e + p) v) = \rho v \cdot b.
\end{aligned} \tag{3.22}$$

Closer inspection of equations (3.22) reveals that there are six unknowns (ρ, v, e and p) and only five equations.Hence, in order to solve the system of equations (3.22), additional constitutive relations must be considered.

The total energy per unit mass e can be written as the summation of internal energy and kinetic energy:

$$e = i + \frac{1}{2} v \cdot v, \tag{3.23}$$

where i is the internal energy. For an ideal gas the internal energy is proportional to the temperature such that

$$i = c_v T, \tag{3.24}$$

where T is the absolute temperature and c_v is the specific heat at constant volume.Up to this point two more equations have been introduced as well as two more unknowns (i and T). The constitutive equation for an ideal gas is used to close the system:

$$p = R \rho T, \tag{3.25}$$

where R is the ideal gas constant per unit mass. Combination of equations(3.23), (3.24) and (3.25) yields

$$e = c_v T + \frac{1}{2} v \cdot v = c_v \frac{p}{R\rho} + \frac{1}{2} v \cdot v. \tag{3.26}$$

R, c_v and c_p are related by

$$c_p = \frac{\gamma R}{\gamma - 1} \text{ and } c_v = \frac{R}{\gamma - 1}, \tag{3.27}$$

where c_p is the specific heat at constant pressure and $\gamma = c_p/c_v$. Substitution of equation (3.27) into (3.26) leads to

$$e = \frac{p}{(\gamma - 1)\rho} + \frac{1}{2} v \cdot v. \tag{3.28}$$

Steady transport problems

This section emphasizes the inefficiencies of the traditional finite element method to solve even the most basic transport problem in one dimension. The steady transport problem consists of a convection-diffusion equation with a source term where derivatives with respect to time are zero and the transport velocity $v = a$ is constantThis type of differential equation describes the behavior of dispersion of chemical pollutants or temperature distribution in flow problems. Considering equation (3.20) under these assumptions and making $e = u$, $\rho v \cdot b = s$, $q = -k \cdot \nabla u$, $\sigma = 0$ one can write

$$a \cdot \nabla u - \nabla \cdot (\nu \nabla u) = s, \tag{3.29}$$

where u is a scalar unknown, a is the convection (or advection) velocity, ν is the diffusivity and s is the source term.Accompanying equation (3.20) two types of boundary condition may appear: Dirichlet (D) and Newmann (N) boundary conditions.

$$u = u_D \quad \text{on } \Gamma_D$$
$$n \cdot (\nu \nabla u) = h \quad \text{on } \Gamma_N. \tag{3.30}$$

The finite element method does not work directly with the differential equation or the strong form. Initially, it is required to write the equivalent weak form of equation (3.30). Hence,considering a trial function u that satisfies the Dirichlet (or essential) boundary conditions and a weighting function w that satisfies the homogeneous Dirichlet boundary conditions [13]:

$$\int_\Omega w(a \cdot \nabla u)d\Omega - \int_\Omega w\nabla \cdot (\nu \nabla u)d\Omega = \int_\Omega wsd\Omega. \tag{3.31}$$

Recall that equation (3.31) is valid for all w that satisfies the homogeneous Dirichlet boundary conditions. Integration by parts of the diffusivity term, and recalling that $w = 0$ on Γ_D,

$$\int_\Omega w(a \cdot \nabla u)d\Omega + \int_\Omega \nabla w \cdot (\nu \nabla u)d\Omega = \int_\Omega wsd\Omega + \int_{\Gamma_N} whd\Gamma, \tag{3.32}$$

where the divergence theorem has been used. It s now convenient to define the operators

$$a(w, u) = \int_\Omega \nabla w \cdot (\nu \nabla u)d\Omega, \quad c(a, w, u) = \int_\Omega w(a \cdot \nabla u)d\Omega$$
$$(w, s) = \int_\Omega wsd\Omega, \quad (w, h)_{\Gamma_N} = \int_{\Gamma_N} whd\Gamma \tag{3.33}$$

in order to write equation (3.32) in the compact form

$$a(w, u) + c(a, w, u) = (w, s) + (w, h)_{\Gamma_N}. \tag{3.34}$$

Equation (3.34) is the starting point for the conventional Galerkin approximation.The trial and weighting functions are written with the aid of shape functions Φ_i defined at specific locations (nodes) of the domain Ω:

$$w \approx \sum_i \Phi_i w_i \quad \text{and } u \approx \sum_j \Phi_j u_j \tag{3.35}$$

Substitution of equations (3.35) into (3.34) leads to

$$\sum_i \sum_j w_i u_j \, a(\Phi_i, \Phi_j) + \sum_i \sum_j w_i u_j \, c(a, \Phi_i, \Phi_j) = \sum_i w_i (\Phi_i, s) + \sum_i w_i (\Phi_i, h)_{\Gamma_N} \tag{3.36}$$

and, since w is arbitrary, so are the w_is, and equation (3.36) transforms into

$$\sum_j a(\Phi_i, \Phi_j) u_j + \sum_j c(a, \Phi_i, \Phi_j) u_j = (\Phi_i, s) + (\Phi_i, h)_{\Gamma_N}, \quad \text{for all i.} \tag{3.37}$$

Following the standard matrix assembly procedure equation (3.37) can be written as

$$(K + C)u = f, \tag{3.38}$$

where K is the stiffness matrix, C is the convective matrix, u is the nodal vector of unknowns and f is the vector of forcing terms. It is a simple matter to identify the corresponding terms between equations (3.37) and (3.38). The stiffness matrix K is related to term a(w, u) defined in equation (3.33) which, in turn, relates to the diffusivity term in the original differential equation (3.29). This is the term also present in structural problems, associated with strain energy and, therefore, to the elliptic character of the transport problem. The convective matrix C is related to term c(a, w, u) defined in equation (3.33) which, in turn, relates to the convective term in the original differential equation (3.29). This is the term that appears because of the Eulerian coordinate system and, therefore, is responsible for the hyperbolic character of the transport problem. The force term f incorporates the contribution of both the source term s and the Newmann boundary conditions.

In order to obtain the stiffness and convective matrices it is necessary to select the shape functions Φ. The simplest shape functions applicable to the one-dimensional steady transport problem are the piecewise linear functions. From the element point of view these shape functions are illustrated in figure 3.4 and they are given by

$$\Phi_1(\xi) = \frac{1-\xi}{2} \quad \text{and} \quad \Phi_2(\xi) = \frac{1+\xi}{2}, \qquad (3.39)$$

where ξ is the local element coordinate, $-1 \leq \xi \leq +1$.

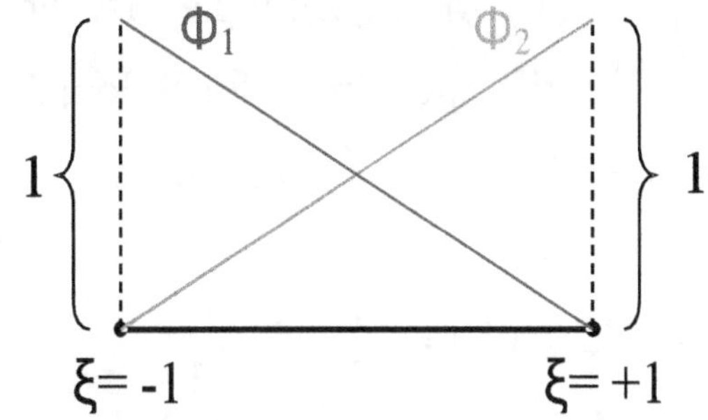

Figure 3.4: Linear shape functions

The element stiffness and convective matrices are obtained when equation (3.39) is substituted into equation (3.37) and the operators defined in equation (3.33) are applied. For an element of length h with a constant coefficient of diffusivity ν and constant advection velocity a these matrices are

$$K_e = \int_{\Omega_e} \nu \begin{bmatrix} \Phi_{1,x}\Phi_{1,x} & \Phi_{1,x}\Phi_{2,x} \\ \Phi_{2,x}\Phi_{1,x} & \Phi_{2,x}\Phi_{2,x} \end{bmatrix} dx = \frac{\nu}{h} \begin{bmatrix} +1 & -1 \\ -1 & +1 \end{bmatrix}$$

$$C_e = \int_{\Omega_e} a \begin{bmatrix} \Phi_1\Phi_{1,x} & \Phi_1\Phi_{2,x} \\ \Phi_2\Phi_{1,x} & \Phi_2\Phi_{2,x} \end{bmatrix} dx = \frac{a}{2} \begin{bmatrix} -1 & +1 \\ -1 & +1 \end{bmatrix}. \qquad (3.40)$$

It is clear that the convective element matrix C_e in equation (3.40) is non-symmetric. Hence,the resulting global convective matrix C in equation (3.38) is also non-symmetric. This calls for the utilization of solvers able to handle non-symmetric matrices that require more memory when compared to the symmetric matrix solvers that store only half the matrix.

At this juncture it is interesting to define the P´eclet number Pe:

$$Pe = \frac{ah}{2\nu}. \qquad (3.41)$$

The importance of the P´ecletnumber lies in the fact that it is a measureof the relative contribution betweendiffusion (ν) and convection(a). The more convectiondominatedthe problem is the higher is the Peclet number.

The deficienciesof the conventional Galerkin schemeto solve convection dominated problems becomes apparent by solving a simple case where the source term in the model equation (3.29) is $s = 1$ and the one-dimensional domain has total length $L = 1$. Exact solution of this problem for $u(0) = u(1) = 0$ is

$$u(x) = \frac{1}{a}\left(x - \frac{1 - \exp(\gamma x)}{1 - \exp(\gamma)}\right), \qquad (3.42)$$

where $\gamma = a/\nu$. When solving the same problem numerically using ten elements, $a = 1$ and for different P´eclet numbers (0.25, 0.9 and 5.0) one obtains the results shown in figure 3.5 where the dashed lines represent the exact solution given by equation (3.42) and the solid lines represent the numerical solution.

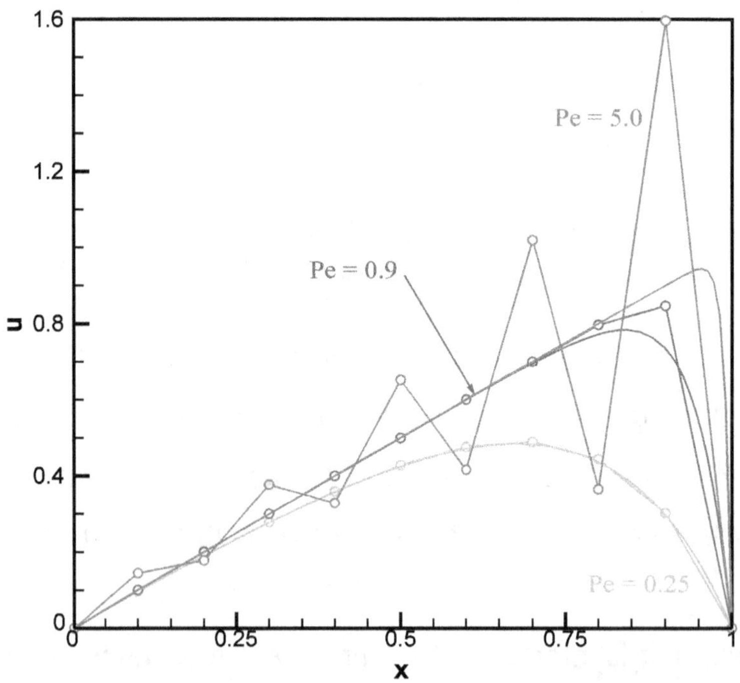

Figure 3.5: Deficiencies of the traditional Galerkin method

The 1D SUPG method

It is obvious from figure 3.5 that wide oscillations occur when the transport is dominated by convection mechanismsi,.e., large P´eclet numbers.Increasing the number of elements in the FE mesh certainly alleviates the numerical problem but it cannot be completely eliminated by that subterfuge alone.This can be seen considering the effect of the number of elements on equation (3.41).More elements lead to smaller h,which, in turn, leads to a smaller P´eclet number.However,one of the most attractive features of the FE method is its ability to approximate solutions with a relatively small number of simple elements. Within this context, mesh refinement enhances the results but tends to overburden the numerical procedures.

The question now is why the conventional Galerkin method does not deliver accurate results when the Peclet number is large. The answer appears by comparing the exact solution of the problem, equation (3.42), against the finite element solution. Considering a uniform FE mesh and two adjacent elements as shown in figure 3.6 it is possible to write the stencil of the finite element scheme associated with line i:

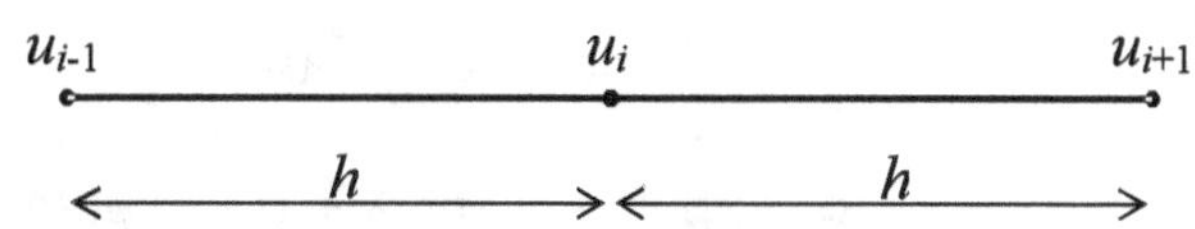

Figure 3.6: Uniform 1D FE mesh

$$\frac{v}{h^2}(-u_{i-1}+2u_i-u_{i+1})+\frac{a}{2h}(u_{i+1}-u_{i-1})=1. \tag{3.43}$$

Equation (3.43) can be recast into another form if equation (3.41) is considered:

$$\frac{a}{2h}\quad\frac{Pe-1}{Pe}u_{i+1}+\frac{2}{Pe}u_i-\frac{Pe+1}{Pe}u_{i-1}=1. \tag{3.44}$$

In order to compare the exact and the numerical solutions it is important to determine what should have been the FE scheme stencil to precisely reproduce the exact solution given by equation (3.42). Assuming that node i has coordinate x_i and using equation (3.42),

$$u_{i-1} = u(x_i - h) = \frac{1}{a}\left[x_i - h - \frac{1-\exp(\gamma(x_i - h))}{1-\exp(\gamma)}\right]$$

$$u_i = u(x_i) = \frac{1}{a}\left[x_i - \frac{1-\exp(\gamma x_i)}{1-\exp(\gamma)}\right]$$

$$u_{i+1} = u(x_i + h) = \frac{1}{a}\left[x_i + h - \frac{1-\exp(\gamma(x_i + h))}{1-\exp(\gamma)}\right]. \qquad (3.45)$$

The stencil presented in equation (3.44) is of the form

$$\alpha_1 u_{i-1} + \alpha_2 u_i + \alpha_3 u_{i+1} = 1, \qquad (3.46)$$

where α_1, α_2, α_3 are coefficients to be determined. Substitution of equations (3.45) into (3.46) and grouping terms that are independent of x_i, linear in x_i and linear in $\exp(x_i)$, the following conditions must be met:

$$\alpha_1 + \alpha_2 + \alpha_3 = 0$$

$$\alpha_3 - \alpha_1 = a/h$$

$$\alpha_1 \exp(-\gamma h) + \alpha_2 + \alpha_3 \exp(\gamma h) = 0. \qquad (3.47)$$

Solving equations (3.47) for α_1, α_2, α_3

$$\alpha_1 = \frac{a}{h}\left[\frac{\exp(\gamma h) - 1}{2 - \exp(\gamma h) - \exp(-\gamma h)}\right] = \frac{a}{h}\left[\frac{1}{\exp(-\gamma h) - 1}\right] = -\frac{a}{2h}(1 + \coth Pe)$$

$$\alpha_2 = \frac{a}{h}\coth Pe$$

$$\alpha_3 = \frac{a}{h}\left[\frac{1 - \exp(-\gamma h)}{2 - \exp(\gamma h) - \exp(-\gamma h)}\right] = \frac{a}{h}\left[\frac{\exp(-\gamma h)}{\exp(-\gamma h) - 1}\right] = \frac{a}{2h}(1 - \coth Pe). \qquad (3.48)$$

Hence, substitution of equation (3.48) into (3.46) yields

$$\frac{a}{2h}\left[(1 - \coth Pe)u_{i+1} + (2\coth Pe)u_i - (1 + \coth Pe)u_{i-1}\right] = 1, \qquad (3.49)$$

Comparison betweenequations (3.49) and (3.44) revealsthe error incurred when the numerical FE solution is adopted: coth(Pe) is approximated by Pe^{-1}. If a parameter β is definedas

$$\beta = \coth Pe - \frac{1}{Pe} \tag{3.50}$$

then equation (3.49) can be re-written as

$$a\frac{u_{i+1} - u_{i-1}}{2h} - (\nu + \tilde{\nu})\frac{u_{i+1} - 2u_i + u_{i-1}}{h^2} = 1, \tag{3.51}$$

where $\tilde{\nu} = \beta a h/2$, or

$$\frac{1-\beta}{2}\, a\frac{u_{i+1} - u_i}{h} + \frac{1+\beta}{2}\, a\frac{u_i - u_{i-1}}{h} - \nu\frac{u_{i+1} - 2u_i + u_{i-1}}{h^2} = 1. \tag{3.52}$$

Parameter $\tilde{\nu}$ present in equation (3.51) represents an artificial diffusivity whereas equation (3.52) shows that different weights are applied to the convective terms, either upwind $(u_i - u_{i-1})$ or downwind $(u_{i+1} - u_i)$. From equation (3.50) it is clear that β tends to zero as $Pe \to 0$ since, in this case,

$$\beta \approx \frac{Pe}{3} + \frac{(Pe)^3}{30} + \ldots. \tag{3.53}$$

Equation (3.53) is the basis for the definition of a simpler expression for β:

$$\beta \approx \begin{cases} -1 & \text{if} \quad Pe \leq -3 \\ Pe/3 & \text{if} \quad -3 < Pe < 3 \\ +1 & \text{if} \quad Pe \geq 3. \end{cases} \tag{3.54}$$

The motivation for the use of approximation (3.54) is that no exponential functions must be evaluated what speeds up computations.The curves for both expressions for β and its approximated value are shown in figure 3.7.

From the discussion just presented it is concluded that the numericalinstabilities can be overcome by numericalschemes that result in stencils of either form (3.51) or (3.52).

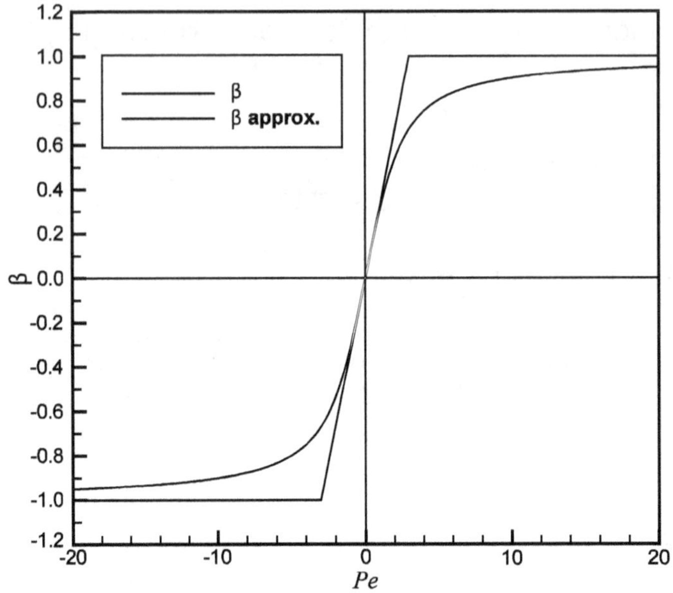

Figure 3.7: β parameter

One of the most successful of such schemes is the streamline upwind Petrov–Galerkin method (SUPG). Consider equation (3.29) and define the differential operator L:

$$L(u) = \mathbf{a} \cdot \nabla u - \nabla \cdot (\nu \nabla u) - s. \tag{3.55}$$

The SUPG stabilization technique consists in writing the weak form of the differential equation (3.29) as

$$a(w, u) + c(\mathbf{a}, w, u) + \underbrace{\sum_e \int_{\Omega^e} (\mathbf{a} \cdot \nabla w)\,\tau\, L(u)\,d\Omega}_{\text{stabilization term}} = (w, s) + (w, h)_{\Gamma_N}, \tag{3.56}$$

where it is understood that w and u are discrete approximations to the respective exact solutions and τ is the stabilization parameter defined as

$$\tau = \frac{\bar{\nu}}{\mathbf{a} \cdot \mathbf{a}}, \tag{3.57}$$

with $\bar{\nu}$ given by $\beta a h/2$ for 1D problems. Notice that the stabilization term in equation

(3.56) tends to zero when the differential equation is satisfied because,in this situation, $L(u) \equiv 0$. When a 1D problem with no Newmann boundary conditions ($h = 0$) equation (3.56) reducesto:

$$\int_0^L \nu w_{,x} u_{,x} dx + \int_0^L a w u_{,x} dx + \sum_e \int_0^{h_e} (a w_{,x}) \tau (a u_{,x} - \nu u_{,xx} - s) dx = \int_0^L s w dx. \quad (3.58)$$

If linear elementsare employedthen $u_{,xx} \equiv 0$ and equation (3.58) becomes

$$\sum_e \left(\int_0^{h_e} \nu w_{,x} u_{,x} dx + \int_0^{h_e} a w u_{,x} dx + \int_0^{h_e} a^2 \tau w_{,x} u_{,x} dx \right) =$$

$$\sum_e \left(\int_0^{h_e} s w dx + \int_0^{h_e} a \tau s w_{,x} dx \right). \quad (3.59)$$

Substitution of equation (3.57) into (3.59) and re-arrangingterms,

$$\sum_e \left[\int_0^{h_e} (\nu + \underline{\nu}) w_{,x} u_{,x} dx + \int_0^{h_e} a w u_{,x} dx = \right] \sum_e \int_0^{h_e} (w + a\tau w_{,x}) s dx. \quad (3.60)$$

The left hand side of equation (3.60) is readily related to the stencil in equation (3.51) with the artificial diffusivity term $\underline{\nu}$. However, the right hand side doesnot correspond to the usual loading term. Another interpretation to (3.59) may be provided if the re-arrangementof terms is madedifferently:

$$\sum_e \left[\int_0^{h_e} \nu w_{,x} u_{,x} dx + \int_0^{h_e} a (w + a\tau w_{,x}) u_{,x} dx = \right] \sum_e \int_0^{h_e} (w + a\tau w_{,x}) s dx. \quad (3.61)$$

The weighting function

$$\tilde{w} = w + a\tau w_{,x} = w + \frac{\beta h}{2} w_{,x} \quad (3.62)$$

can be interpreted as a modified weighting function. In the traditional Galerkin scheme the weighting functions are shown in equations (3.39). However,the modified weighting function $\tilde{w}$ for the 1D piecewise linear element is discontinuous as illustrated in figure 3.8, where $\tilde{N}_i$ is obtained by substitution of w by N_i given in equation (3.39). As shown in

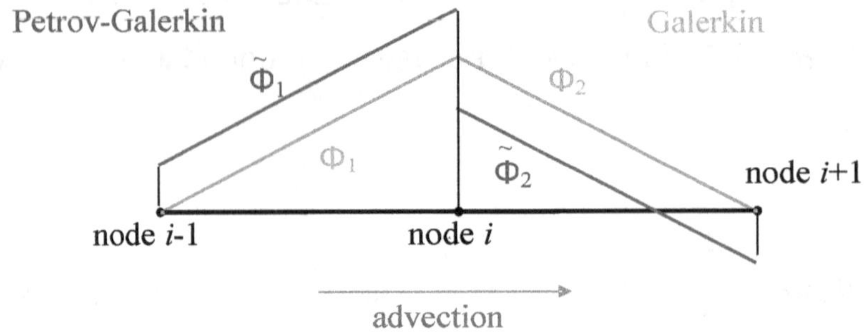

Figure 3.8: Petrov-Galerkin weighting function

equation (3.61) the modified weighting function$\tilde{w}$ must be used for the convective term as well as the loading term.

Figure 3.8 shows a positive advection velocityHence, it can be seen that the convective term upwind (to the left of node i) is more strongly weighted than the convective term downwind (to the right of node i). Therefore, the denomination "streamline upwind" comes from this observation whereas the denomination "Petrov-Galerkin" comes from the fact that different weighting functions are used for u and w.

The discretization scheme proposed in equation (3.61) can be used to investigate, again, the convective-diffusive problem of equation (3.29) when s = 1, L = 1 and u(0) = u(1) = 0. The results are displayed in figure 3.9. Comparison with figure 3.5 shows that the deficiencies of the traditional Galerkin method are overcome when the streamline upwind Petrov-Galerkin (SUPG) method is used.

The multidimensional SUPG method

The correction parameter β defined in equation (3.50) was obtained when a one-dimensional problem was analyzed.Extension of the SUPG method to two- or three-dimensional problems in not straightforward. The difficulty is that the artificial diffusivity should be added only in the direction of the flow and not transversely.When crosswind artificial diffusivity is added the problem becomes overly diffusive and inaccurate results are delivered.

Hughes [25]proposed a method to add artificial diffusivity only in the flow direction and not crosswind.A simple strategy to select v in equation (3.57) is to make

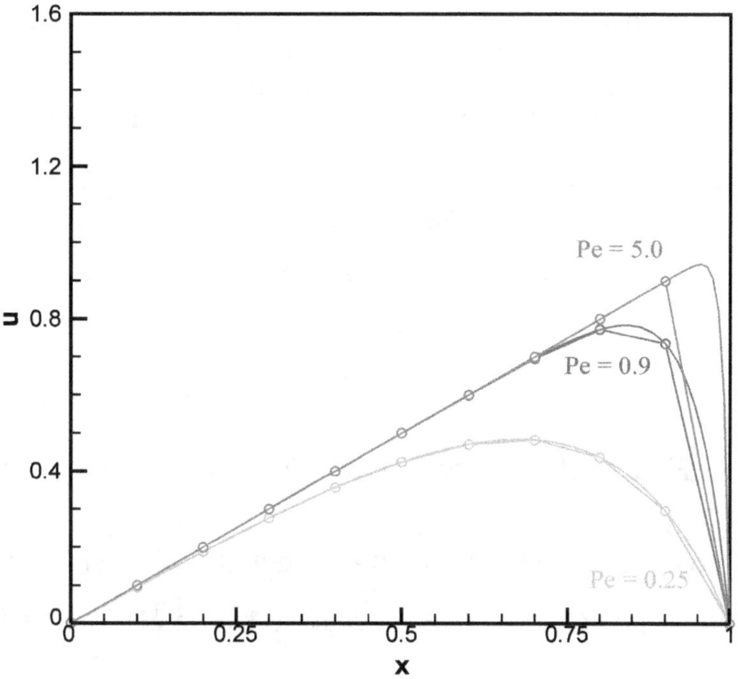

Figure 3.9: 1D convective-diffusive problem:SUPG results

$$\bar{\nu} = \frac{1}{2}\left(\bar{\xi} a_\xi h_\xi + \bar{\eta} a_\eta h_\eta\right) , \tag{3.63}$$

where

$$\begin{aligned} \bar{\xi} &= \coth \mathbf{P}e_\xi - 1/\mathbf{P}e_\xi & \bar{\eta} &= \coth \mathbf{P}e_\eta - 1/\mathbf{P}e_\eta \\ Pe_\xi &= a_\xi h_\xi/(2\nu) & Pe_\eta &= a_\eta h_\eta/(2\nu) \\ a_\xi &= e_\xi \cdot a & a_\eta &= e_\eta \cdot a \end{aligned} \tag{3.64}$$

and versors e_ξ, e_η and lengths h_ξ, h_η are defined in figure 3.10 that illustrates a 2D bi-linear quadrilateral element. Observe that the stabilization scheme proposed in equations (3.63) and (3.64) is not unique.In fact, there are other ways to stabilize multidimensional problems [26].

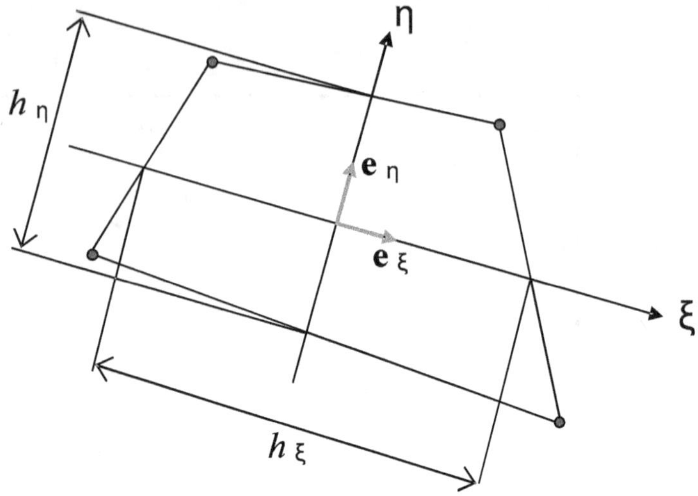

Figure 3.10: 2D convective-diffusive problem parameters

Operators a(w, u) and c(a, w, u) defined in equation (3.33) and the stabilization term introduced in equation (3.56) can be specialized to the 2D case as follows:

$$a(w,u) = \int_\Omega \nabla w \cdot (\nu \nabla u) d\Omega = \int_\Omega \nu (w_{,x} u_{,x} + w_{,y} u_{,y}) d\Omega$$

$$c(a,w,u) = \int_\Omega w (a \cdot \nabla u) d\Omega = \int_\Omega w (a_x u_{,x} + a_y u_{,y}) d\Omega$$

$$\int_{\Omega_e} (a \cdot \nabla w) \tau L(u) d\Omega = \int_{\Omega_e} (a_x u_{,x} + a_y u_{,y}) \tau (a_x u_{,x} + a_y u_{,y} - \nu u_{,xx} - \nu u_{,yy} - s) d\Omega, \quad (3.65)$$

where a_x is the x component of the advection velocity and a_y is its y component.

Discretization of the above expressions and terms (w, s) and (w_{Γ_N}, h) in equation (3.33) can be done with the aid of equations (3.35). Similar to the structural problem, if matrix Φ is defined as in equation (2.12) to store the shape functions ϕ_i then the element matrices become:

$$K_e = \int_{\Omega_e} \nu (\Phi^T_{,x} \Phi_{,x} + \Phi^T_{,y} \Phi_{,y}) d\Omega$$

$$C_e = \int_{\Omega_e} (a_x \Phi^T \Phi_{,x} + a_y \Phi^T \Phi_{,y}) d\Omega$$

$$f_e = \int_{\Omega_e} \Phi^T s d\Omega + \int_{\Gamma_N \cap \partial\Omega_e} \Phi^T h d\Gamma - \int_{\Omega_e} (a_x \Phi^T_{,x} + a_y \Phi^T_{,y}) \tau s d\Omega$$

$$S_e = \int_{\Omega_e} (a_x \Phi^T_{,x} + a_y \Phi^T_{,y}) \tau (a_x \Phi_{,x} + a_y \Phi_{,y} - \nu \Phi_{,xx} - \nu \Phi_{,yy}) d\Omega, \quad (3.66)$$

where $\partial\Omega_e$ is the boundary of element e, K_e is the element stiffness matrix, C_e is the elementconvectivematrix, f_e is the elementload vector and S_e is the elementstabilization matrix.

Notice that it is necessaryto compute$\Phi_{i,x}$ and $\Phi_{i,y}$ in equation(3.66) what can be done as in equation (2.7). Moreover, it is also necessaryto compute secondderivatives with respectto x and y. In the 1D casethe term $u_{,xx}$ in the stabilization term is identically zero sinceΦ_1 and Φ_2 are both linear functions of ξ. However,evenwhen bilinear elementsare used, the shapefunctions contain $\xi\eta$ products that do not in generalallow $\Phi_{i,xx}$, $\Phi_{i,yy}$ or $\Phi_{i,xx}$ to vanishidentically. Thesecanbe evaluatedif the following relationship is considered:

$$\begin{vmatrix} \Phi_{i,\xi\xi} \\ \Phi_{i,\xi\eta} \end{vmatrix} = \begin{vmatrix} x_{,\xi}^2 & 2x_{,\xi}y_{,\xi} & y_{,\xi}^2 \\ x_{,\xi}x_{,\eta} & x_{,\xi}y_{,\eta} + x_{,\eta}x_{,\xi} & y_{,\xi}y_{,\eta} \end{vmatrix} \begin{vmatrix} \Phi_{i,xx} \\ \Phi_{i,xy} \end{vmatrix} + \begin{vmatrix} \Phi_{i,x}x_{,\xi\xi} + \Phi_{i,y}y_{,\xi\xi} \\ \Phi_{i,x}x_{,\xi\eta} + \Phi_{i,y}y_{,\xi\eta} \end{vmatrix} . \quad (3.67)$$

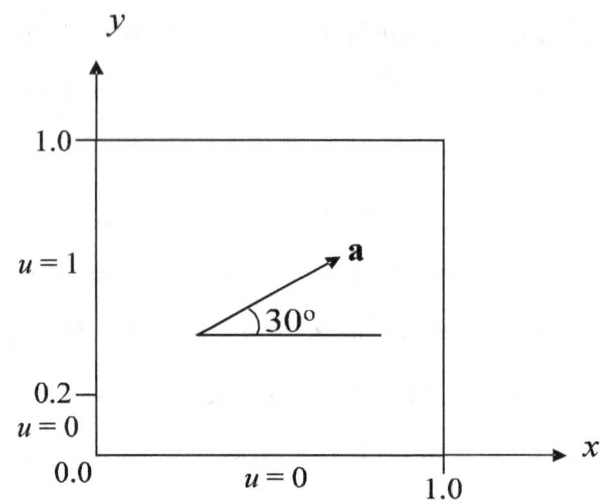

Figure 3.11: Advection skew to the mesh

diffusion is negligible and the differential governing equation is essentially hyperbolic.

Overshoot and undershoot can be seen in figure 3.12 in the vicinity of the shockT.his is a numerical inaccuracy that can be improved either by mesh refinement (notice that the 10×10 mesh is relatively coarse) or by speciabperators purposedly designed to capture shocks.

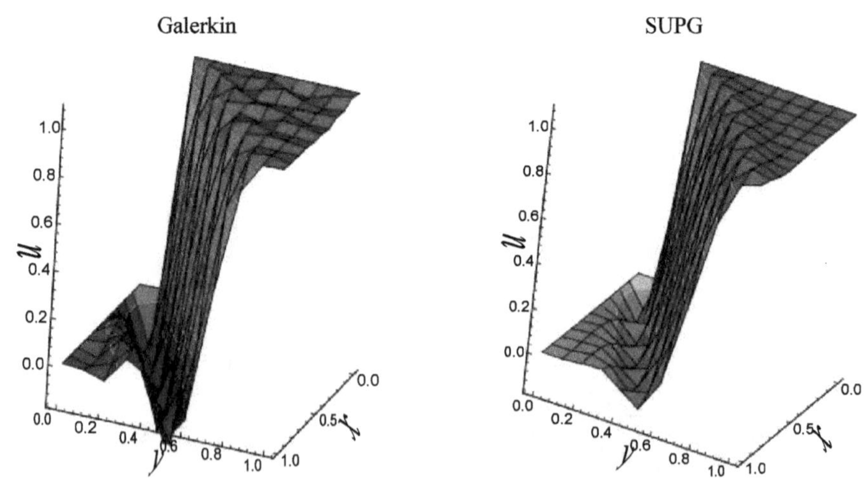

Figure 3.12: 2D problem with homogeneous Newmann outflow boundary condition

Consider now the same problem of figure 3.11 but, instead of homogeneous Newmann condition in the outflow portion of the boundary, assume a homogeneous Dirichlet Newmann condition in the outflow portion of the boundary. The results for this case are presentedin figure 3.13 where totally inaccurate values are obtained by the traditional Galerkin method (observethe disparity in the z axis scales).

The valid SUPG result shows,again, a shock due to the discontinuousboundary condition. It is also possible to see the presence of a boundary layer along the outflow edges of the domain. The boundary layer is too thin to be accurately described by the coarse 10×10 mesh.Hence, mesh refinement, at least along the outflow edges, should be done in order to enhance numerical results.

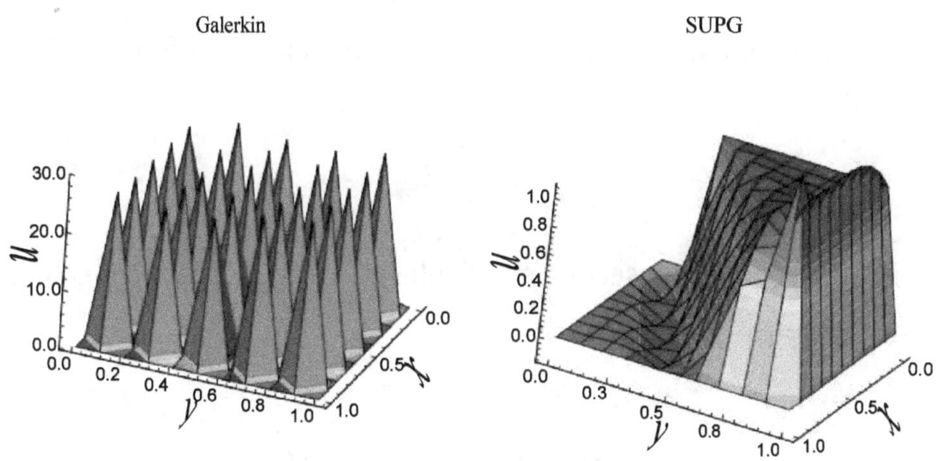

Figure 3.13: 2D problem with homogeneous Dirichlet outflow boundary condition

Compressible flow problems

The transport problem previously investigated has two fundamental simplifications embedded: it is assumed to be steady, i.e., time independent and the advection velocity is constant. However,diffusive effects are present since the term $\nabla \cdot \sigma$ in equation (3.15) or $\nabla \cdot q$ in equation (3.20) are retained.These simplifications are applicable to a very limited class of problem and, therefore, it is of great interest to relax those simplifications in order to enlarge the universe of phenomena amenable to analytical and quantitative treatment.

A common assumption in compressible flow problems is that the viscous effects can be neglected ($\sigma = 0$) and the problem is adiabatic ($\nabla \cdot q = 0$). These assumptions lead to the formulation of the Euler equationsin compressiblefluid dynamics presentedin equation (3.22). If the fluid velocity v is not known, unlike the transport problem, then the convectiveterms in equation(3.22) are nonlinearand somekind of numerical nonlinear solver must be used. Moreover, becausethe variables are time dependent,all derivatives with respect to time must be taken into account what further complicatesthe numerical procedures.

3.5.1 Euler equations revisited

The 2D Euler equationscan be more compactly written in the form

$$\frac{\partial u}{\partial t} + \frac{\partial f_x}{\partial x} + \frac{\partial f_y}{\partial y} = s \tag{3.68}$$

where u is the vector of conservative variables, f_x and f_y are flux vectors and s is the load ve These vectors are given by:

$$u = \begin{vmatrix} \rho \\ \rho u \\ \rho v \\ \rho e \end{vmatrix}, \quad f_x = \begin{vmatrix} \rho u \\ \rho u^2 + p \\ \rho uv \\ (\rho e + p)u \end{vmatrix}, \quad f_y = \begin{vmatrix} \rho v \\ \rho uv \\ \rho v^2 + p \\ (\rho e + p)v \end{vmatrix} \quad \text{and } s = \begin{vmatrix} 0 \\ \rho b \\ \rho v \cdot b \end{vmatrix}. \tag{3.69}$$

The derivatives of the flux vectors present in equation (3.68) can expressed as

$$\frac{\partial f_x}{\partial x} = \frac{\partial f_x}{\partial u}\frac{\partial u}{\partial x} = A_x \frac{\partial u}{\partial x} \quad \text{and} \quad \frac{\partial f_y}{\partial y} = \frac{\partial f_y}{\partial u}\frac{\partial u}{\partial y} = A_y \frac{\partial u}{\partial y}, \tag{3.70}$$

where, through consideration of equation (3.28), matrices A_x and A_y are computed as follows [27]:

$$A_x = \begin{vmatrix} 0 & 1 & 0 & 0 \\ (\gamma - 1)q^2/2 - u^2 & (3 - \gamma)u & (1 - \gamma)v & \gamma - 1 \\ -uv & v & u & 0 \\ (\gamma - 1)q^2u/2 - uh & h - (\gamma - 1)u^2 & (1 - \gamma)uv & \gamma u \end{vmatrix}$$

$$
A_y = \begin{vmatrix} 0 & 0 & 1 & 0 \\ -uv & v & u & 0 \\ (\gamma-1)q^2 - v^2 & (1-\gamma)u & (3-\gamma)v & \gamma - 1 \\ (\gamma-1)vq^2/2 - vh & (1-\gamma)uv & h - (\gamma-1)v^2 & \gamma v \end{vmatrix}, \qquad (3.71)
$$

where $q^2 = \mathbf{v} \cdot \mathbf{v}$ and h being the total entalpy defined as

$$
h = e + \frac{p}{\rho} = \frac{\gamma p}{\rho(\gamma-1)} + \frac{1}{2}q^2. \qquad (3.72)
$$

At this point it is possible to define the Mach number as

$$
M = \frac{q}{c}, \qquad (3.73)
$$

where c is the speed of sound given by

$$
c = \sqrt{\frac{\gamma p}{\rho}}. \qquad (3.74)
$$

Substitution of equations (3.70) into (3.68) yields

$$
u_{,t} + A_x u_{,x} + A_y u_{,y} = s, \qquad (3.75)
$$

where it must be highlighted that both A_x and A_y depend on the conservative variables in vector u. Equation (3.75) is known as the quasi linear form of the Euler equations.

The Euler differential equation obtained must be solved along with the applicable boundary conditions. However,specification of boundary conditions is not a trivial task [28]. Perhaps the greatest difficulty is that the appropriate boundary conditions depend on the flow and a wrong choice could lead to nonunique solutions or even to no viable solutions.

Since the Euler equations do not assume diffusive effects, i.e., viscous effects, a solid wall can exert force only along its normal. The nonpenetrating condition states that the fluid velocity normal to the solid wall must be zero, however,neither density nor total energy can be specifiedat a solid wall. On the other hand massand energyflux at the solid wall is zero but theseconditions are automatically taken care of by the finite elementmethod.

A rigorous treatment of boundary conditions in the domain contour where there is no solid wall involvesthe investigation of the eigenvaluesof

$$\det(I\lambda - k_x A_x - k_y A_y) = 0, \qquad (3.76)$$

where k_x, k_y are real value wave numbers [23]. In summary, the applicable boundary conditions depend on the local flow regime:

1. Supersonic ($M > 1$)
   - Inflow boundary ($v_n < 0$): all components of u must be specified
   - Outflow boundary ($v_n > 0$): no component of u is specified
2. Subsonic ($M < 1$)
   - Inflow boundary ($v_n < 0$): three components of u must be specified (2D flow)
   - Outflow boundary ($v_n > 0$): only one component of u may be specified

where v_n is the velocity component normal to the boundary.

FE formulation

Assuming that the gravity effects are negligible, i.e., $b = 0$, the finite element formulation of the Euler equations can be written as [29]:

$$\int_\Omega w^T(u_{,t} + A_x u_{,x} + A_y u_{,y})d\Omega + \sum_e \int_{\Omega_e} \tau(A_x w_{,x} + A_y w_{,y})^T(u_{,t} + A_x u_{,x} + A_y u_{,y})d\Omega + \sum_e \int_{\Omega_e} \delta(w^T_{,x} u_{,x} + w^T_{,y} u_{,y})d\Omega = 0, \qquad (3.77)$$

where w is an arbitrary vector of weighting functions satisfying the homogeneous Dirichlet boundary conditions, τ is the stabilization parameter and δ is a shock capturing parameter.

Considerableresearcheffort hasbeendevotedto determining both the stabilization parameter τ and the shock capturing parameter δ [30, 31, 32, 33]. The necessityof the SUPG stabilization parameter is evident from the previous one- and two-dimensionalexamples previously presented.The shock capturing parameter is important to eliminate or reduce the numerical over and undershootobserved,for instance,in figure 3.12.

A commonstrategy to determine the stabilization parameter τ is, again, basedon the wave equation (3.76) and the spectral radius of A_x and A_y. Working with the Euler equationsin the nonconservativeform (ρ, u, v and p as variables)it can be shown[27] that the eigenvaluesof equation (3.76) are

$$\begin{aligned}\lambda_1 = \lambda_2 &= k_x u + k_y v \\ \lambda_3 &= k_x u + k_y v + c(k_x^2 + k_y^2)^{1/2} \\ \lambda_4 &= k_x u + k_y v - c(k_x^2 + k_y^2)^{1/2}\end{aligned} \tag{3.78}$$

where c is the speed of sound given in equation (3.74)Equation (3.78) permits the computation of the spectral radii of A_x (when $k_x = 1, k_y = 0$) or A_y (when $k_x = 0, k_y = 1$). These are given by:

$$\lambda_x = \max(|u|, |u + c|, |u - c|) \quad \text{and} \quad \lambda_y = \max(|v|, |v + c|, |v - c|). \tag{3.79}$$

The directional stabilization parameters τ_x and τ_y are defined as

$$\tau_x = \frac{\alpha h_x}{\lambda_x} \quad \text{and} \quad \tau_y = \frac{\alpha h_y}{\lambda_y}, \tag{3.80}$$

where α is a free parameter usually taken to be 0.5 and h_x and h_y are the directional element sides

$$h_x = 2\sqrt{x_{,\xi}^2 + x_{,\eta}^2} \quad \text{and} \quad h_y = 2\sqrt{y_{,\xi}^2 + y_{,\eta}^2}. \tag{3.81}$$

The stabilization parameter used in equation (3.77) is the maximum of τ_x or τ_y. Determination of the shock capturing parameter δ is more elaborate and involves the inverse of a matrix A_0 that arises in the study of the symmetrized form of the Euler equations[30]. Firstly, it is convenient to define auxiliary variables ι, ϑ, v_1, v_2, v_3, v_4, φ as

$$\iota = \rho e - \frac{q}{2\rho}, \quad \vartheta = \ln\left[\frac{(\gamma - 1)\iota}{\rho^{\gamma}}\right],$$

$$v_1 = 1 + \gamma - \vartheta - \frac{\rho e}{\iota}, \quad v_2 = \frac{\rho u}{\iota}, \quad v_3 = \frac{\rho v}{\iota}, \quad v_4 = -\frac{\rho}{\iota}, \quad \varphi = \frac{v_2^2 + v_3^2}{2 v_4}. \tag{3.82}$$

The inverse of A_0 is obtained as

$$A_0^{-1} = -\frac{1}{\iota v_4}\begin{bmatrix} \varphi^2 + \gamma & \varphi v_2 & \varphi v_3 & (\varphi + 1)v_4 \\ \varphi v_2 & v_2^2 - v_4 & v_2 v_3 & v_2 v_4 \\ \varphi v_3 & v_2 v_3 & v_3^2 - v_4 & v_3 v_4 \\ (\varphi + 1)v_4 & v_2 v_4 & v_3 v_4 & v_4^2 \end{bmatrix}. \tag{3.83}$$

Finally, parameter δ is

$$\delta = \left[\frac{(A_x u_{,x} + A_y u_{,y})^T A_0^{-1} (A_x u_{,x} + A_y u_{,y})}{(\xi_{,x} u_{,x} + \xi_{,y} u_{,y})^T A_0^{-1} (\xi_{,x} u_{,x} + \xi_{,y} u_{,y}) + (\eta_{,x} u_{,x} + \eta_{,y} u_{,y})^T A_0^{-1} (\eta_{,x} u_{,x} + \eta_{,y} u_{,y})}\right]^{1/2} \tag{3.84}$$

Differently from the steady transport problem, the differential equation (3.77) is time dependent and, therefore, some kind of numerical discretization in time must be used. The time derivative is replaced by

$$u_{,t} \approx \frac{\Delta u}{\Delta t} = \frac{u(t^{n+1}) - u(t^{n})}{t^{n+1} - t^{n}}, \tag{3.85}$$

where $\Delta t = t^{n+1} - t^{n}$ is a suitable time step. Different ways of computing $\Delta u/\Delta t$ result in different time discretization schemes. The simplest scheme consists in replacing the time derivatives $u_{,t}$ by its forward finite difference equivalent

$$\frac{u^{n+1} - u^{n}}{\Delta t} = u_{,t}^{n} \tag{3.86}$$

This method is fully explicit since,if the flow variablesare known at time t^n, then they may be obtained at time t^{n+1} through solution of a linear system of algebraic equations. Implicit schemesdo not presentthis nice feature becausethe systemof algebraicequations fails to be linear and somekind of iterative proceduremust be called upon.

The disadvantageof the schemeproposed in equation (3.86) is that it is only first order accurate, what usually meansthat extremely small time stepsΔt are required. An alternative to improve the order of accuracyis the two-step Taylor-Galerkin schemewhere an intermediate flow state is calculatedbetweentimes t^n and t^{n+1}. This is doneasfollows:

$$u^{n+1/2} = u^n + \frac{\Delta t}{2} u^n_{,t}$$

$$u^{n+1} = u^n + \Delta u^{n+1/2}_{,t}, \tag{3.87}$$

and the weakform shownin equation (3.77) is applied to the secondstep of the procedure. In the first step $u^{n+1/2}$ is evaluatedusing equation (3.75) for $s = 0$.

$$u^{n+1/2} = u^n + \frac{\Delta t}{2} u^n_{,t} = u^n - \frac{\Delta t}{2}\left(A^n_x u^n_{,x} + A^n_y u^n_{,y}\right). \tag{3.88}$$

Notice that matrices A_x and A_y dependon u and, therefore,they are alsotime dependent. Spacederivatives of u are required in equation (3.88). They are usually obtained through computation of u at nodal points and, subsequently,using spacederivativesof the shapefunctions $\Phi_{,x}$ and $\Phi_{,y}$. Equation (3.77) is now employedin the secondstep of the two-step Taylor-Galerkin scheme:

$$\int_\Omega w^T\left(\frac{\Delta u}{\Delta t} + A_x^{n+1/2} u^{n+1/2}_{,x} + A_y^{n+1/2} u^{n+1/2}_{,y}\right) d\Omega +$$

$$\sum_e \int_{\Omega_e} \tau \left(A_x^{n+1/2} w_{,x} + A_y^{n+1/2} w_{,y}\right)^T \left(\frac{\Delta u}{\Delta t} + A_x^{n+1/2} u^{n+1/2}_{,x} + A_y^{n+1/2} u^{n+1/2}_{,y}\right) d\Omega +$$

$$\sum_e \int_{\Omega_e} \delta\left(w^T_{,x} u^{n+1/2}_{,x} + w^T_{,y} u^{n+1/2}_{,y}\right) d\Omega = 0. \tag{3.89}$$

Vectors u and w can discretized in space as

$$u = \Phi q_e, \quad w = \Phi q_{we}, \quad \Delta u = \Phi \Delta q_e, \tag{3.90}$$

where q_{we} are arbitrary nodal variables. Substitution of equations(3.90) into (3.89) yields

$$\sum_e \int_{\Omega_e} q_{we}^T \Phi^T \left(\Phi \frac{\Delta q_e}{\Delta t} + A_x^{n+1/2} u_{,x}^{n+1/2} + A_y^{n+1/2} u_{,y}^{n+1/2} \right) d\Omega +$$

$$\sum_e \int_{\Omega_e} \tau \left(A_x^{n+1/2} \Phi q_{,x\,we} + A_y^{n+1/2} \Phi q_{,y\,we} \right)^T \left(\Phi \frac{\Delta q_e}{\Delta t} + A_x^{n+1/2} u_{,x}^{n+1/2} + A_y^{n+1/2} u_{,y}^{n+1/2} \right) d\Omega +$$

$$\sum_e \int_{\Omega_e} \delta (q_{we}^T \Phi_{,x}^T u_{,x}^{n+1/2} + q_{we}^T \Phi_{,y}^T u_{,y}^{n+1/2}) d\Omega = 0. \qquad (3.91)$$

Notice that all vectorsand matricescontaining a superscriptn+1/2 are known from the first step of the two-step Taylor-Galerkin scheme.Consideringthe arbitrariness of vector q_{we} an element mass matrix and load vector can be defined as

$$M_e = \int_{\Omega_e} (\Phi + \tau A_x^{n+1/2} \Phi_{,x} + \tau A_y^{n+1/2} \Phi_{,y})^T \Phi d\Omega$$

$$f_e = -\int_{\Omega_e} (\Phi + \tau A_x^{n+1/2} \Phi_{,x} + \tau A_y^{n+1/2} \Phi_{,y})^T (A_x^{n+1/2} u_{,x}^{n+1/2} + A_y^{n+1/2} u_{,y}^{n+1/2}) d\Omega$$

$$- \int_{\Omega_e} \delta (\Phi_{,x}^T u_{,x}^{n+1/2} + \Phi_{,y}^T u_{,y}^{n+1/2}) d\Omega. \qquad (3.92)$$

According to equations (3.91) and (3.92) the vector of nodal variables is Δq_e and not q_e. Hence, once Δq_e^n is found, q_e is updated by $q_e^{n+1} = q_e^n + \Delta q_e^n$.

The stabilization term successfully eliminates the spurious numerical oscillations observed in the traditional Galerkin method. However, it can be seen in equation (3.92) that it leads to a nonsymmetric mass matrix and, worse, to a time dependant mass matrix. This is a highly undesirable feature of the SUPG method applied to gas dynamics problems that was not perceived in the steady transport problem because time dependency was not an issue.

The time dependency of the mass matrix means that it changes from time t^n to time t^{n+1}. Therefore, in order to solve for Δq in $M \Delta q = f$, the mass matrix must be decomposed at every time. However,it is expected that M^{n+1} should not be drastically different from M^n what suggests that, if sufficiently small time steps are used,q^{n+1} and q^n should not be too different either. This observation suggests that some kind of iterative solver for the system of algebraic equations could be employed where the knowledge ofq^n would be helpful to obtain q^{n+1}. The iterative solver used in the present project is the GCR (generalized conjugateresidual) describedin appendix B.

An issueof relevanceis the imposition of solid wall boundary conditions. Sincethere is one cartesian globalreference system xy the velocity components u and v are aligned along the x and y axes respectively.However, there may be situations when the solid wall is not oriented parallel to the global reference system.Consider figure 3.14 that shows a two-dimensional situation where the solid wall is oriented at an angle θ with respect to the global x axis.

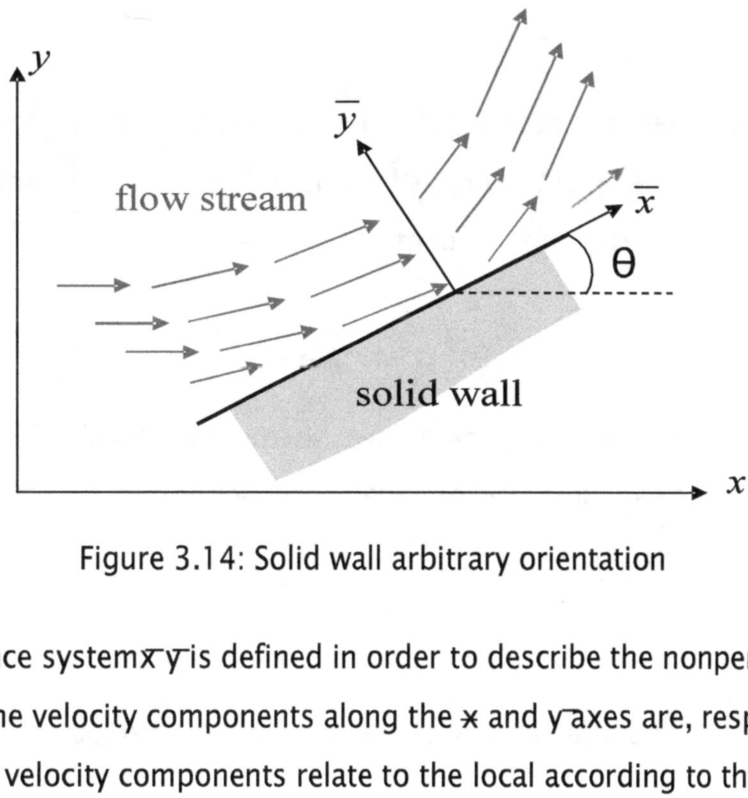

Figure 3.14: Solid wall arbitrary orientation

A local reference system $\bar{x}\bar{y}$ is defined in order to describe the nonpenetrating condition according to it. The velocity components along the $\bar{x}$ and $\bar{y}$ axes are, respectively $\bar{u}$, and $\bar{v}$. The local global velocity components relate to the local according to the transformation

$$\begin{Bmatrix} u \\ v \end{Bmatrix} = \begin{bmatrix} \cos\theta & -\sin\theta \\ \sin\theta & \cos\theta \end{bmatrix} \begin{Bmatrix} \bar{u} \\ \bar{v} \end{Bmatrix}. \quad (3.93)$$

The nonpenetrating condition requires that $\bar{v} = 0$. However, element matrices and vectors were obtained considering the globalvelocity components u and v. Therefore, in order to enforce the solid wall condition, a transformation matrix must be used to go from the global to the local reference system whenever requiredI.nspired in equation (3.93) a global transformation matrix is definedas

$$T_i = \begin{array}{c} \begin{matrix} & & 4i-2 & 4i-1 & & \\ & & \downarrow & \downarrow & & \end{matrix} \\ \begin{bmatrix} 1 & & & & & \\ & \ddots & & & & \\ & & \cos\theta & -\sin\theta & & \\ & & \sin\theta & \cos\theta & & \\ & & & & \ddots & \\ & & & & & 1 \end{bmatrix} \end{array}, \qquad (3.94)$$

where it is assumed that node i is on a solid wall. The system of algebraic equations obtained when the element matrices in equation (3.92) are assembled into global matrices at time t^n can be put in the form

$$Mq = f, \qquad (3.95)$$

where vector q contains components described in the global reference system. Assuming that node i must have its velocity variables described in a local coordinate system equation (3.95) can be written as

$$T_i^T M T_i \bar{q} = T_i^T f, \qquad (3.96)$$

where $q = T_i \bar{q}$. It should be obvious that the matrix products $T_i^T M T_i$ and $T_i^T f$ are not computed by direct multiplications since T_i contains severalzero entries. In fact, it is possible to write

$$\overline{M} = T_i^T M T_i = \begin{bmatrix} \cdots & \cdots & \cdots & \overline{m}_{1,4i-2} & \overline{m}_{1,4i-1} & \cdots & \cdots \\ \cdots & \cdots & \cdots & \overline{m}_{2,4i-2} & \overline{m}_{2,4i-1} & \cdots & \cdots \\ & & & \vdots & \vdots & & \\ \overline{m}_{4i-2,1} & \overline{m}_{4i-2,2} & \cdots & \overline{m}_{4i-2,4i-2} & \overline{m}_{4i-2,4i-1} & \cdots & \overline{m}_{4i-2,n} \\ \overline{m}_{4i-1,1} & \overline{m}_{4i-1,2} & \cdots & \overline{m}_{4i-1,4i-2} & \overline{m}_{4i-1,4i-1} & \cdots & \overline{m}_{4i-1,n} \\ & & & \vdots & \vdots & & \\ \cdots & \cdots & \cdots & \overline{m}_{n,4i-2} & \overline{m}_{n,4i-1} & \cdots & \cdots \end{bmatrix}, \quad (3.97)$$

where

$$\begin{aligned}
\overline{m}_{k,4i-2} &= m_{k,4i-2}\cos\theta + m_{k,4i-1}\sin\theta, \quad \text{for } k < 4i-2,\ k > 4i-1 \\
\overline{m}_{k,4i-1} &= -m_{k,4i-2}\sin\theta + m_{k,4i-1}\cos\theta, \quad \text{for } k < 4i-2,\ k > 4i-1 \\
\overline{m}_{4i-2,k} &= m_{4i-2,k}\cos\theta + m_{4i-1,k}\sin\theta, \quad \text{for } k < 4i-2,\ k > 4i-1 \\
\overline{m}_{4i-1,k} &= -m_{4i-2,k}\sin\theta + m_{4i-1,k}\cos\theta, \quad \text{for } k < 4i-2,\ k > 4i-1 \\
\overline{m}_{4i-2,4i-2} &= m_{4i-2,4i-2}\cos^2\theta + m_{4i-1,4i-1}\sin^2\theta + (m_{4i-2,4i-1} + m_{4i-1,4i-2})\sin\theta\cos\theta \\
\overline{m}_{4i-2,4i-1} &= m_{4i-2,4i-1}\cos^2\theta - m_{4i-1,4i-2}\sin^2\theta + (m_{4i-1,4i-1} - m_{4i-2,4i-2})\sin\theta\cos\theta \\
\overline{m}_{4i-1,4i-2} &= m_{4i-1,4i-2}\cos^2\theta - m_{4i-2,4i-1}\sin^2\theta + (m_{4i-1,4i-1} - m_{4i-2,4i-2})\sin\theta\cos\theta \\
\overline{m}_{4i-1,4i-1} &= m_{4i-1,4i-1}\cos^2\theta + m_{4i-2,4i-2}\sin^2\theta - (m_{4i-2,4i-1} + m_{4i-1,4i-2})\sin\theta\cos\theta.
\end{aligned} \quad (3.98)$$

A simple and similar transformation rule can be obtained for the load vector $T_i^T f$. Notice that all the entries of the mass matrix that do not belong to row or column $4i-2$ and $4i-1$ are not affected by the transformation in equation (3.98). It is not unusual to have severalnode on a solid wall. In this case the transformation described in equations (3.97) and (3.98) must applied as many times as required.

Equation (3.96) deals with global matrices.However, the procedures presented in equations (3.97) and (3.98) can be applied at the element level, before the element mass matrix and load vector are stored into the corresponding global arrays. In that case matrix M is replaced by the element mass matrix M_e and vector f is replaced by the element load vector f_e, both presented in equation (3.92).

Numerical results

Three examples are investigated to show the efficacy of the FE CFD code implemented.

Supersonic parallel jet interaction

Two supersonic jets are initially separated by a membrane that is suddenly removed.Both jets mix up forming a shock wave and a Prandt-Meyer expansion wave. The computational domain is $\Omega = [0, 1] \times [0, 1]$ and a 40×40 mesh of bilinear isoparametric elements isemployed.The left edge is the inlet portion of the boundary whereas the other three edgesare assumed to outlets.Figure 3.15 shows the inlet conditions.

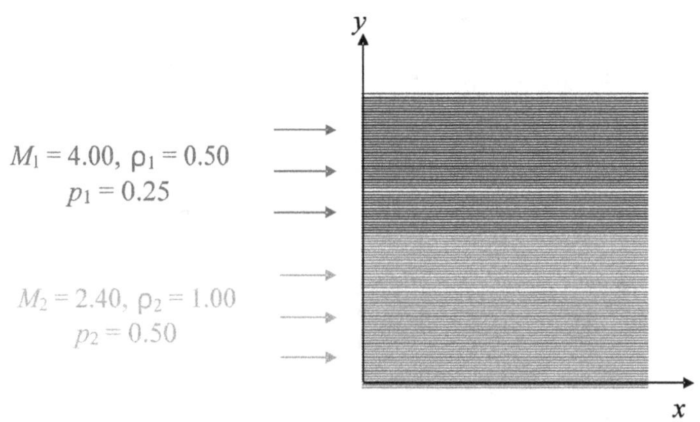

Figure 3.15: Supersonic parallel jets

The Mach number contours are presented in figure 3.16. A shock wave forms in the region where initially only the M = 4.0 flow was present whereas an expansion wave forms in the region where initially only the M = 2.4 flow was present

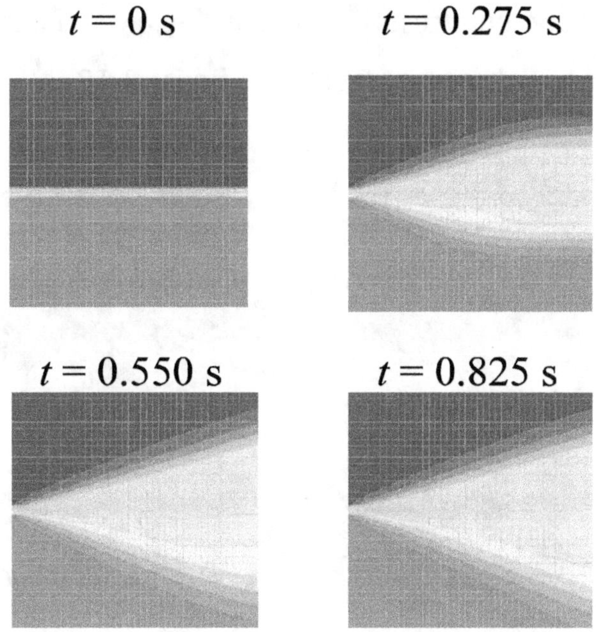

Figure 3.16: Supersonic parallel jets interaction

Supersonic wedge channel

This problem consists of an inlet supersonic flow with Mach number M = 2.0, density ρ = 0.25 and pressure p = 0.25. The channelhas a wedge that must be considered.The imposition of the solid wall boundary condition on the wedge is done according to equations (3.93)-(3.98). A 60 × 20 mesh is used as shown in figure 3.17.

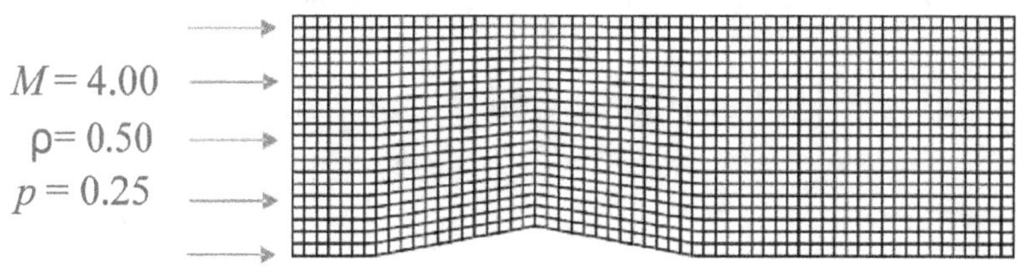

Figure 3.17: Supersonic wedge channel mesh

Figure 3.18 shows the Mach contours until a steady state is reached.The wedge induces the formation of both shock and expansion waves.Two shock waves form when the flow turns into itself, i.e., at the beginning and at end of the wedge whereas an expansion wave forms at the apex of the wedge.The first shock wave bounces off of the upper channelwall and bouncesoff, again, at the lower channelwall. Thereafter it mixes up with the secondshockwaveas detailed in figure 3.19.

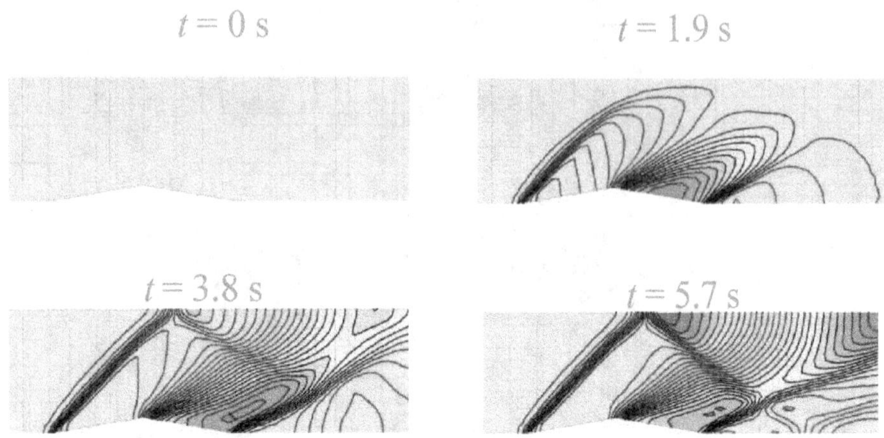

Figure 3.18: Supersonic wedge channel r:esults

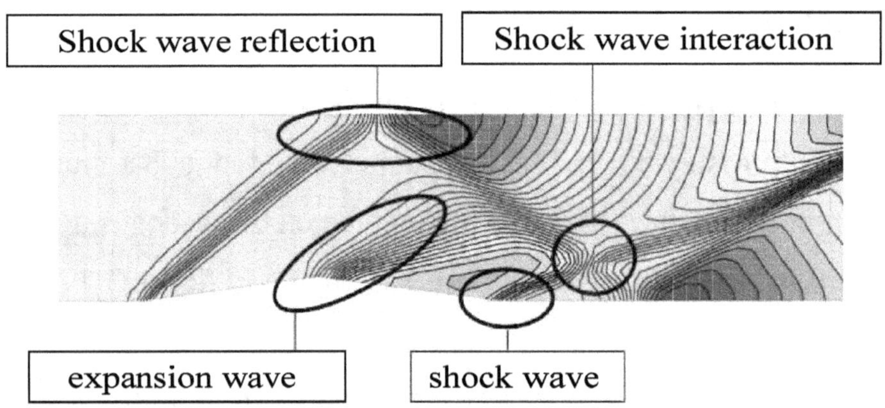

Figure 3.19: Supersonic wedge channel, shock and expansion

waves

Supersonic flow past a cylinder

This problem is perhaps the most challenging of the three because a detached shock wave appears ahead of the cylinder and there is a near expansion behind the cylinder. An elliptic mesh is used to model the domain as seen in figure 3.20.

The results shown in figures 3.21 and 3.22 reveal that most of the detached shock has

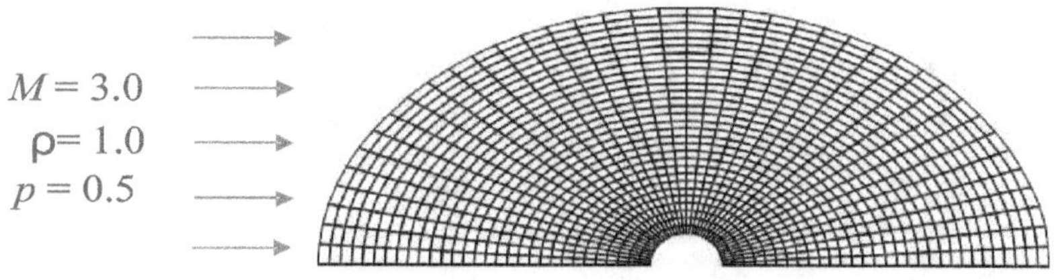

Figure 3.20: Supersonic flow past a cylinder:mesh

been captured within two elements.Moreover, the distance from the shock to the cylinder is in good agreement with experimental tests [29].

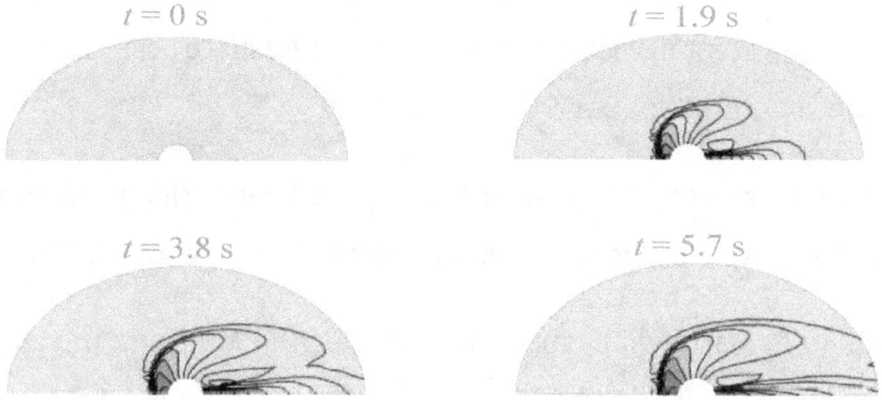

Figure 3.21: Supersonic flow past a cylinder:results

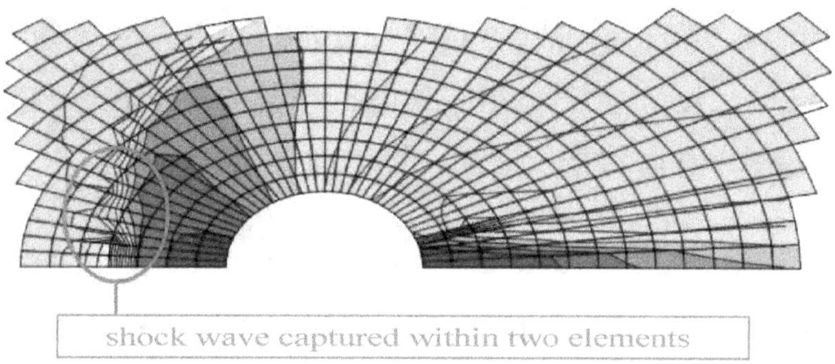

Figure 3.22: Supersonic flow past a cylinder:bow shock wave

Bibliography

[1] SobieskiJS, and Haftka RT (1996) Multidisciplinary aerospace design optimization: survey of recent developments.34th AIAA Aerospace Sciences Meeting and Exhibit, Reno, NV, EUA, paper 96-0711.

[2] Alonso JJ, Jameson A (1994) Fully-implicit time-marching aeroelastic solutions. 32nd AIAA Aerospace Sciences Meeting and Exhibit, Reno, NV, EUA, paper 94-0056.

[3] Alosno JJ, Martinelli L, and Jameson A (1995) Multigrid unsteady Navier-Stokes calculations with aeroelastic applications. 33rd AIAA Aerospace Sciences Meeting and Exhibit, Reno, NV, EUA, paper 95-0048.

[4] Wakayama S, and Kroo IM (1990) A method for lifting surface design using nonlinear optimization. AIAA/AHS/ASEE Aircraft Design Systems Conference, Dayton, OH, EUA, paper 90-3120.

[5] Gallman JW (1992) Structural and Aerodynamic Optimization of Joined-Wing Aircraft. PhD dissertation, Department of Aeronautics and Astronautics,Stanford University, Stanford, CA, USA.

[6] Zienkiewicz OC, and Taylor RL (1991) The Finite Element Method. 4th edition, vols. 1 & 2, London: McGraw-Hill.

[7] Tannehill JC, Anderson CA, and Pletcher RH (1997) Computational Fluid Dynamics and Heat Transfer. 2nd edition, New York, Taylor & Francis.

Example, $u_I = 1$, $u_J = 0$.

$$\begin{Bmatrix} f_I \\ f_J \end{Bmatrix} = k\begin{bmatrix} 1 & -1 \\ -1 & 1 \end{bmatrix}\begin{Bmatrix} u_I = 1 \\ u_J = 0 \end{Bmatrix} = \begin{Bmatrix} k \\ -k \end{Bmatrix}$$

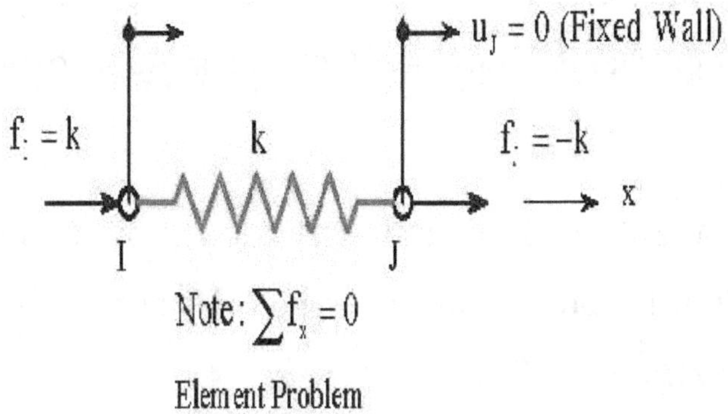

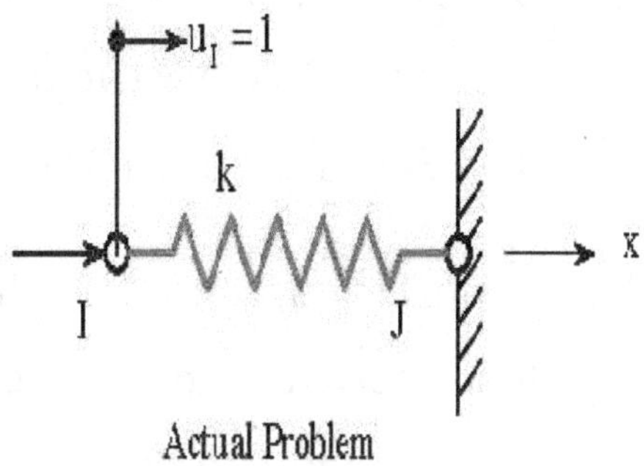

Spring element has <u>one</u> rigid body mode.

Inter-Element Axial Displacement

The axial displacement (u) is continuous through the assembled mesh and is described by a linear polynomial within each element. Each element in the mesh may be described by a different linear polynomial, depending on the spring rate (k), external loading, and constraints on the element.

Inter-Element Deformation

The deformation (Δ) is piecewise constant through the assembled mesh and is described by a constant within each element. Each element in the mesh may be described by a different constant, depending on the spring constant (k), external loading, and constraints on the element.

Inter-Element Internal Axial Force

The internal axial force (f) is piecewise continuous through the assembled mesh and is described by a constant within each element. Each element in the mesh may be described by a different constant, depending on the spring constant, external loading, and constraints on the element.

[8] Reuther JJ, Alonso JJ, Martins JRRA, and Smith SC. A coupled aero-structural optimization method for complete aircraft configurations. 37th AIAA Aerospace Sciences Meeting and Exhibit, Reno, NV, EUA, paper 99-0187.

[9] Deblois BM (1996) Quadratic, streamline upwinding for finite element method solutions to 2-D convective transport problems. Computer Methods in Applied Mechanics and Engineering, 134(2):107-15.

[10] Sampaio PAB, and Coutinho ALGA (2001) A natural derivation of discontinuity capturing operator for convection-diffusion problems. Computer Methods in Applied Mechanics and Engineering, 190(47):6291-308.

[11] Turner MJ, Clough, RW, and Martin HC (1956) Stiffness and deflection analysis of complex structures. Journalof AeronauticalSociety, 32(9): 805-823.

[12] Clough RW (1960) The finite element method in plane stress analysis. Proc. 2nd ASCE Conference on Electronic Computation, Pittsburg, PA, Sept. 8-9.

[13] Hughes TJR (1987) The Finite Element Method. Prentice-Hall Inc., Englewood Cliffs, New Jersey.

[14] Zienkiewicz OC (1989) The Finite Element Method. 4th edition, vol. 1 and 2, McGraw-Hill, Singapore.

[15] Bathe K-J, and Wilson EL (1976) Numerical Methods in Finite Element Analysis. Prentice Hall Inc., New Jersey.

[16] Fried I, and Yang SK (1973) Triangular nine-degree-of-freedom, C^0 plate bending elements of quadratic accuracy. Quarterly Journal of Applied Mechanics, 3310–312.

[17] Garnet H, Crouzet-Pascal J, and Pifko AB (1980) Aspects of a simple triangular plate bending finite element. Computers and Structures, 12(6)7: 83789.

[28] Donea J, Huerta A (2003) Finite Element Methods for Flow Problems. Chichester, West Sussex, John Wiley & Sons.

[29] Le Beau GJ, Tezduyar TE (1991) Finite element computation of compressible flows with the SUPG formulation. Advances in Finite Element Analysis and Dynamics, FED-Vol. 123, ASME, New York, pp. 21-27.

[30] Hughes TJR, Franca LP, Mallet M (1986) A new finite element formulation for computational fluid dynamics: I. symmetric forms of the compressible Euler and Navier-Stokes and the second law of thermodynamics. Computer Methods in Applied Mechanics and Engineering, 54(2):223-234.

[31] Hughes TJR, Mallet M, Mizukami A (1986) A new finite element formulation for computational fluid dynamics: II. beyond SUPG. Computer Methods in Applied Mechanics and Engineering, 54(3):341-355.

[32] Hughes TJR, Mallet M (1986) A new finite element formulation for computational fluid dynamics: III. the generalized streamline operator for multidimensional advective-diffusive systems.Computer Methods in Applied Mechanics and Engineering,58(3): 305-328.

[33] Hughes TJR, Mallet M (1986) A new finite element formulation for computational fluid dynamics: IV. a discontinuity capturing operator for multidimensional advective-diffusive systems.Computer Methods in Applied Mechanics and Engineering,58(3): 329-336.

[34] Rosen R (1968) Matrix bandwidth minimization. Proceedings of the 23rd Association for Compuating Machinery National Conference, pp. 585-595.

[35] Cuthill EH, and McKee JM (1969) Reducing the bandwidth of sparse symmetric matrices. Proceedings of the 24th Association for Compuating Machinery NationaCl onference., pp. 151-172.

[36] Saad Y (2003) Iterative Methods for Sparse Linear Systems. Philadelphia, SIAM, 2nd edition.

QUESTION BANK

PART A QUESTIONS WITH ANSWERS

UNIT 1

1. What is meant by finite element?

   A small units having definite shape of geometry and nodes is called finite element.

2. What is meant by node or joint?

   Each kind of finite element has a specific structural shape and is inter- connected with the adjacent element by nodal point or nodes. At the nodes, degrees of freedom are located. The forces will act only at nodes at any others place in the element.

3. What is the basic of finite element method?

   Discretization is the basis of finite element method. The art of subdividing a structure in to convenient number of smaller components is known as discretization.

4. What are the types of boundary conditions?

   Primary boundary conditions

   Secondary boundary conditions

5. State the methods of engineering analysis?

   Experimental methods

   Analytical methods

   Numerical methods or approximate methods

6. What are the types of element?

7. 1D element

   2D element

   3D element

8. State the three phases of finite element method.

   Preprocessing

   Analysis

   Post Processing

9. What is structural problem?

   Displacement at each nodal point is obtained. By these displacements solution stress and strain in each element can be calculated.

10. What is non structural problem?

Temperature or fluid pressure at each nodal point is obtained. By using these values properties such as heat flow fluid flow for each element can be calculated.

10. What are the methods are generally associated with the finite element analysis?

Force method

Displacement or stiffness method.

11. Explain stiffness method.

Displacement or stiffness method, displacement of the nodes is considered as the unknown of the problem. Among them two approaches, displacement method is desirable.

12. What is meant by post processing?

Analysis and evaluation of the solution result is referred to as post processing. Postprocessor computer program help the user to interpret the result by displaying them in graphical form.

13. Name the variation methods.

Ritz method.

Ray-Leigh Ritz method.

14. What is meant by degrees of freedom?

When the force or reaction act at nodal point node is subjected to deformation. The deformation includes displacement rotation, and or strains. These are collectively known as degrees of freedom

15. What is meant by discretization and assemblage?

The art of subdividing a structure in to convenient number of smaller components is known as discretization. These smaller components are then put together. The process of uniting the various elements together is called assemblage.

16. What is Rayleigh-Ritz method?

It is integral approach method which is useful for solving complex structural problem, encountered in finite element analysis. This method is possible only if a suitable function is available.

17. What is Aspect ratio?

It is defined as the ratio of the largest dimension of the element to the smallest dimension. In many cases, as the aspect ratio increases the in accuracy of the solution increases. The conclusion of many researches is that the aspect ratio

18. What is truss element?

The truss elements are the part of a truss structure linked together by point joint which transmits only axial force to the element.

19. What are the h and p versions of finite element method?

It is used to improve the accuracy of the finite element method. In h version, the order of polynomial approximation for all elements is kept constant and the numbers of elements are increased. In p version, the numbers of elements are maintained constant and the order of polynomial approximation of element is increased.

20. Name the weighted residual method

Point collocation method

Sub domain collocation method

Least squares method

Galerkins method.

UNIT 2

21. List the two advantages of post processing.

Required result can be obtained in graphical form. Contour diagrams can be used to understand the solution easily and quickly.

22. During discretization, mention the places where it is necessary to place a node?

Concentrated load acting point

Cross-section changing point

Different material interjections

Sudden change in point load

23. What is the difference between static and dynamic analysis?

Static analysis: The solution of the problem does not vary with time is known as static analysis

Example: stress analysis on a beam

Dynamic analysis: The solution of the problem varies with time is known as dynamic analysis

24. Name any four FEA softwares.

ANSYS

NASTRAN

COSMOS

25. Differentiate between global and local axes.

Local axes are established in an element. Since it is in the element level, they change with the change in orientation of the element. The direction differs from element to element.

Global axes are defined for the entire system. They are same in direction for all the elements even though the elements are differently oriented.

26. Distinguish between potential energy function and potential energy functional

If a system has finite number of degree of freedom (q_1,q_2,and q_3), then the potential energy expressed as,

$\pi = f(q_1, q_2, \text{and } q_3)$

It is known as function. If a system has infinite degrees of freedom then the potential energy is expressed as

$$\int f\left(x, y, \frac{dy}{dx}, \frac{d^2y}{dx^2} \ldots\ldots\right) dx$$

27. What are the types of loading acting on the structure?

Body force (f)

Traction force (T)

Point load (P)

28. Define the body force

A body force is distributed force acting on every elemental volume of the body Unit: Force per unit volume.

Example: Self weight due to gravity

29. Define traction force

Traction force is defined as distributed force acting on the surface of the body. Unit: Force per unit area.

Example: Frictional resistance, viscous drag, surface shear

30. What is point load?

Point load is force acting at a particular point which causes displacement.

31. What are the basic steps involved in the finite element modeling.

Discretization of structure.

Numbering of nodes.

32. Write down the general finite element equation.

$$\{F\}=[K]\{u\}$$

33. What is discretization?

The art of subdividing a structure in to a convenient number of smaller components is known as discretization.

34. What are the classifications of coordinates?

Global coordinates

Local coordinates

Natural coordinates

35. What is Global coordinates?

The points in the entire structure are defined using coordinates system is known as global coordinate system.

36. What is natural coordinates?

A natural coordinate system is used to define any point inside the element by a set of dimensionless number whose magnitude never exceeds unity. This system is very useful in assembling of stiffness matrices.

37. Define shape function.

Approximate relation $\varphi(x,y) = N_1(x,y)\,\varphi_1 + N_2(x,y)\,\varphi_2 + N_3(x,y)\,\varphi_3$

Where φ_1, φ_2, and φ_3 are the values of the field variable at the nodes N_1, N_2, and N_3 are the interpolation functions.

N_1, N_2, and N_3 are also called shape functions because they are used to express the geometry or shape of the element.

38. What are the characteristic of shape function?

It has unit value at one nodal point and zero value at other nodal points. The sum of shape function is equal to one.

39. Why polynomials are generally used as shape function?

Differentiation and integration of polynomial are quit easy.

The accuracy of the result can be improved by increasing the order of the polynomial. It is easy to formulate and computerize the finite element equations

40. How do you calculate the size of the global stiffness matrix?

Global stiffness matrix size = Number of nodes X Degrees of freedom per node

UNIT 3

41. Write down the expression of stiffness matrix for one dimensional bar element.

$$[K]=\frac{AE}{l}\begin{pmatrix}1 & -1\\ -1 & 1\end{pmatrix}$$

42. State the properties of stiffness matrix

It is a symmetric matrix

The sum of elements in any column must be equal to zero
It is an unstable element. So the determinant is equal to zero.

43. Write down the expression of stiffness matrix for a truss element.

$$\begin{Bmatrix}u_1\\ u_2\end{Bmatrix}=\begin{bmatrix}l & m & 0 & 0\\ 0 & 0 & l & m\end{bmatrix}\begin{Bmatrix}u_1\\ u_2\\ u_3\\ u_4\end{Bmatrix}$$

44. Write down the expression of shape function N and displacement u for one dimensional bar element.

$U=N_1u_1+N_2u_2$

$N_1=1-X/l$

$N_2=X/l$

45. Define total potential energy.

Total potential energy, π = Strain energy (U) + potential energy of the external forces (W)

46. State the principle of minimum potential energy.

Among all the displacement equations that satisfied internal compatibility and the boundary condition those that also satisfy the equation of equilibrium make the potential energy a minimum is a stable system.

47. Write down the finite element equation for one dimensional two noded bar element.

$$[K]=\frac{AE}{l}\begin{bmatrix}1 & -1\\ -1 & 1\end{bmatrix}$$

48. What is truss?

A truss is defined as a structure made up of several bars, riveted or welded together.

49. States the assumption are made while finding the forces in a truss.

All the members are pin jointed.

The truss is loaded only at the joint

The self weight of the members is neglected unless stated.

50. State the principles of virtual energy?

A body is in equilibrium if the internal virtual work equals the external virtual work for the every kinematically admissible displacement field

51. What is essential boundary condition?

Primary boundary condition or EBC Boundary condition which in terms of field variable is known as Primary boundary condition.

52. Natural boundary conditions?

Secondary boundary natural boundary conditions which are in the differential form of field variable is known as secondary boundary condition

53. How do you define two dimensional elements?

Two dimensional elements are define by three or more nodes in a two dimensional plane. The basic element useful for two dimensional analysis is the triangular element.

54. What is CST element?

Three noded triangular elements are known as CST. It has six unknown displacement degrees of freedom (u1, v1, u2, v2, u3, v3). The element is called CST because it has a constant strain throughout it.

55. What is LST element?

Six nodded triangular elements are known as LST. It has twelve unknown displacement degrees of freedom. The displacement function for the elements are quadratic instead of linear as in the CST.

56. What is QST element?

Ten nodded triangular elements are known as Quadratic strain triangle. It is also called as cubic displacement triangle.

58. What meant by plane stress analysis?

Plane stress is defined to be a state of stress in which the normal stress and shear stress directed perpendicular to the plane are assumed to be zero.

60. Define plane strain analysis.

Plane strain is defined to be state of strain normal to the xy plane and the shear strains are assumed to be zero.

UNIT 4

61. Write down the stiffness matrix equation for two dimensional CST elements.

Stiffness matrix $[K]=[B]^T[D][B]At$

$[B]^T$-Strain displacement $[D]$-Stress strain matrix $[B]$-Strain displacement matrix

62. Write down the stress strain relationship matrix for plane stress conditions.

$$\frac{E}{(1+v)(1-2v)}\begin{pmatrix} 1-v & v & 0 \\ 0 & 1-v & 0 \\ 0 & 0 & \frac{1-2v}{2} \end{pmatrix}$$

63. What is axisymmetric element?

Many three dimensional problem in engineering exhibit symmetry about an axis of rotation such type of problem are solved by special two dimensional element called the axisymmetric element

64. What are the conditions for a problem to be axisymmetric?

The problem domain must be symmetric about the axis of revolution All boundary condition must be symmetric about the axis of revolution All loading condition must be symmetric about the axis of revolution

65. Give the stiffness matrix equation for an axisymmetric triangular element. Stiffness matrix $[K]=[B]^T[D][B]2\pi rA$

66. What is the purpose of Isoparametric element?

It is difficult to represent the curved boundaries by straight edges finite elements. A large number of finite elements may be used to obtain reasonable resemblance between original body and the assemblage.

67. Write down the shape functions for 4 noded rectangular elements using natural coordinate system.

$$N_1 = \frac{1}{4}(1-\varepsilon)(1-\eta) \qquad N_2 = \frac{1}{4}(1+\varepsilon)(1-\eta)$$

$$N_3 = \frac{1}{4}(1+\varepsilon)(1+\eta) \qquad N_4 = \frac{1}{4}(1-\varepsilon)(1+\eta)$$

68. Write down Jacobian matrix for 4 noded quadrilateral elements.

$$[J] = \begin{pmatrix} J_{11} & J_{12} \\ J_{21} & J_{22} \end{pmatrix}$$

69. Write down stiffnes matrix equation for 4 noded isoparametric quadrilateral elements.

Stiffness matrix $[K] = t\int_{-1}^{1}\int_{-1}^{1} [B]^T[D][B]\,|J|\,\partial\varepsilon\,\partial\eta$

70. Define super parametric element.

If the number of nodes used for defining the geometry is more than of nodes used for defining the displacement is known as super parametric element

71. Define sub parametric element.

If the number of nodes used for defining the geometry is less than number of nodes used for defining the displacement is known as sub parametric element.

72. What is meant by Isoparametric element?

If the number of nodes used for defining the geometry is same as number of nodes used for defining the displacement is known as Isoparametric element.

73. Is beam element an Isoparametric element?

Beam element is not an Isoparametric element since the geometry and displacement are defined by different order interpretation functions.

74. What is the difference between natural coordinate and simple natural coordinate?

L1 = 1-x/l

L2 = x/l

75. What is Area coordinates?

L1 = A1/A L2 = A2/A L3 = A3/A

76. What is simple natural coordinate?

A simple natural coordinate is one whose value between -1 and 1.

77. Give example for essential boundary conditions.

The geometry boundary condition are displacement, slope.

78. Give example for non essential boundary conditions.

The natural boundary conditions are bending moment, shear force

79. What is meant by degrees of freedom?

When the force or reaction act at nodal point node is subjected to deformation. The deformation includes displacement rotation, and or strains. These are collectivelyknown as degrees of freedom.

80. What is QST element?

Ten noded triangular elements are known as Quadratic strain triangle.It is alsocalled as cubic displacement triangle.

UNIT 5

81. What meant by plane stress analysis?

Plane stress is defined to be a state of stress in which the normal stress and shearstress directed perpendicular to the plane are assumed to be zero.

82. Define plane strain analysis.

Plane strain is defined to be state of strain normal to the x,y plane and the shearstrains are assumed to be zero.

83. What is truss element?

The truss elements are the part of a truss structure linked together by point jointwhich transmits only axial force to the element.

84. List the two advantages of post processing.

Required result can be obtained in graphical form.

Contour diagrams can be used to understand the solution easily and quickly.

85. What are the h and p versions of finite element method?

It is used to improve the accuracy of the finite element method. In h version, the order of polynomial approximation for all elements is kept constant and the numbers of elements are increased. In p version, the numbers of elements are maintained constant and the order of polynomial approximation of element is increased.

86. During discretization, mention the places where it is necessary

87. to place a node?Concentrated load acting point Cross-section changing point Different material inter junction point Suddenchange in point load

88. What is the difference between static and dynamic analysis?

Static analysis: The solution of the problem does notvary with time is known asstatic analysis

Example: stress analysis on a beam

Dynamic analysis: The solution of the problem varies with time is known asdynamic analysis

Example: vibration analysis problem.

89. What is meant by discretization and assemblage?

The art of subdividing a structure in to convenient number of smaller components is known as discretization. These smaller components are then put together. The process of uniting the various elements together is called assemblage.

90. What is Rayleigh-Ritz method?

It is integral approach method which is useful for solving complex structural problem, encountered in finite element analysis. This method is possible only if asuitable function is available.

91. What is Aspect ratio?

It is defined as the ratio of the largest dimension of the element to the smallestdimension. In many cases, as the aspect ratio increases the in accuracy of the solution increases. The conclusion of many researches is that the aspect ratio should be close to unity as possible.

92. What is essential boundary condition?

Primary boundary condition or EBC, Boundary condition which in terms of fieldvariable is known as Primary boundary condition

93. Natural boundary conditions,

Secondary boundary natural boundary conditions
which are in the differentialform of field variable
is known as secondary boundary condition.

94. How do you define two dimensional elements?

Two dimensional elements are define by three or more nodes in a two dimensional plane. The basic element useful for two dimensional analysis is the triangular element.

95. State the principles of virtual energy?

A body is in equilibrium if the internal virtual work equals the external virtual work for the every kinematically admissible displacement field.

96. What is non-homogeneous form?

When the specified values of dependent variables are non-zero, the boundary conditi said to be non-homogeneous.

97. What is homogeneous form?

When the specified values of dependent variables is zero, the boundary condition are said to be homogeneous.

98. Define initial value problem.

An initial value problem is one in which the dependent variable and possibly is derivatives are specified initially.

99. Define boundary value problem.

A differential equation is said to describe a boundary value problem if the dependent variable and its derivatives are required to take specified values on the boundary.

1. O. C. Zienkiewicz and R. L. Taylor, *The Finite Element Method*, 4th ed (McGraw-Hill, New York, 1989).
2. J. N. Reddy, *An Introduction To The Finite Element Method*, Second Edition ed (McGraw-Hill, New York, 1993).
3. R. D. Cook, *Finite Element Modeling For Stress Analysis* (John Wiley & Sons, Inc., New York, 1995).
4. K. J. Bathe, *Finite Element Procedures* (Prentice Hall, Englewood Cliffs, NJ, 1996).
5. T. R. Chandrupatla and A. D. Belegundu, *Introduction To Finite Elements in Engineering*, 3rd ed (Prentice Hall, Upper Saddle River, NJ, 2002).
6. R. D. Cook, D. S. Malkus, M. E. Plesha, and R. J. Witt, *Concepts and Applications of Finite Element Analysis*, 4th ed (John Wiley & Sons, Inc., New York, 2002).
7. S. Moaveni, *Finite Element Analysis - Theory and Application with ANSYS*, 2nd ed (Prentice-Hall, Upper Saddle River, NJ, 2002).

www.ingramcontent.com/pod-product-compliance
Lightning Source LLC
LaVergne TN
LVHW080049210926
848481LV00011B/3315
* 9 7 9 8 2 0 1 1 1 1 4 6 5 *